Springer Series in Optical Sciences

Founding Editor

H.K.V. Lotsch

Volume 243

Springer Series in Optical Sciences is led by Editor-in-Chief William T. Rhodes, Florida Atlantic University, USA, and provides an expanding selection of research monographs in all major areas of optics:

- lasers and quantum optics
- ultrafast phenomena
- optical spectroscopy techniques
- optoelectronics
- information optics
- applied laser technology
- industrial applications and
- other topics of contemporary interest.

With this broad coverage of topics the series is useful to research scientists and engineers who need up-to-date reference books.

Ivan Andronov

Problems of High Frequency Diffraction by Elongated Bodies

Springer

Ivan Andronov
Department of Physics
St Petersburg University
Petrodvorets, Russia

ISSN 0342-4111 ISSN 1556-1534 (electronic)
Springer Series in Optical Sciences
ISBN 978-981-99-1275-9 ISBN 978-981-99-1276-6 (eBook)
https://doi.org/10.1007/978-981-99-1276-6

This Springer imprint is published by the registered company Springer Nature Singapore Pte Ltd.
The registered company address is: 152 Beach Road, #21-01/04 Gateway East, Singapore 189721, Singapore

Contents

List of Figures

Chapter 1
High-Frequency Diffraction and Elongated Bodies

1.1 High-Frequency Methods

When computing a numerical solution to a diffraction problem, in order to achieve acceptable accuracy, the size of the mesh should be a small fraction of the wavelength. This leads to the fact that the number of unknowns rapidly increases with frequency and numerical methods, especially those using volume mesh become computationally very intensive at high frequencies. Therefore, some problems remain unaffordable even for modern computers.

An alternative to pure numerical methods are the asymptotic approaches, which on the contrary become more accurate at high frequencies. The basic concept is the *ray method* [5, 9, 11, 19, 20]. It allows the wave fields to be described as a sum of contributions carried by rays. The ray method is applicable in domains where rays form a regular set of lines. The method fails near the shadowed part of the surface of a body and in the penumbra. Geometrical optics (ray method) predicts on the surface of the object a current twice larger than the amplitude of the incident magnetic field in the lit zone and zero current in the shadow. To improve this wrong approximation, one introduces a more accurate analysis of the fields.

There are two main approaches to high-frequency diffraction. One is the *canonical problems* approach, in which the diffraction effects are studied on the basis of an explicit solution in the case of a canonical problem which may be obtained from variable separation, Wiener-Hopf method [27], Malyuzhinets technique [24, 25] (see also [7]) or any other method. Then this solution, initially written in the form of an integral or an infinite series, is subjected to some kind of transformation, which enables the saddle point method to be applied to the resulting expression. Watson transform [17] appears useful in many such cases. The asymptotic expression obtained by passing these steps is finally rewritten in terms of the general characteristics of the surface. Then if one needs to have a general form expression for an arbitrary surface possessing similar ray picture, the following nonrigorous step of the procedure is done. According to it, the asymptotic expression is assumed to be

I. Andronov, *Problems of High Frequency Diffraction by Elongated Bodies*, Springer Series in Optical Sciences 243, https://doi.org/10.1007/978-981-99-1276-6_1

valid if the parameters of the surface which are usually constant for the canonical problem are replaced with the values of the variable characteristics of the surface taken at a particular point.

The other approach is much more rigorous, and the asymptotic expansions resulting from it can be often justified mathematically. It uses the canonical problem only as a hint, if needed, to invent the form of the asymptotic expression, so-called *Ansatz*. This *Ansatz* is then substituted into the Maxwell or Helmholtz equations and the boundary conditions. If the Ansatz is correctly chosen, this leads to a recurrent system of much simpler problems, which as a result are explicitly solvable. These are the boundary-value problems for the unknown functions presented in the *Ansatz*. Generally speaking, to make the procedure mathematically rigorous, one needs to justify that the resulting expression indeed represents the asymptotic expansion of the solution. However, this step is often omitted by saying that the rigorousity of derivations is maintained at the level usually accepted in Physics.

As was said before, the ray method or quasi-optics *Ansatz* can be used to describe high-frequency fields in domains where rays form a regular set of lines and it fails near the shadowed part of the surface, in penumbra, near caustics or focal points, etc. All the domains, where the ray method is inapplicable, are small in one, two or all three dimensions. This observation enables the approach of the boundary-layer method to provide local asymptotic expansions for the fields in each of such small domains. The basic concepts of the boundary-layer method in diffraction theory can be already found in the works of V. A. Fock and M. A. Leontovich in the 1940s (see, for example, [13–15, 22, 23]) and in the paper of R. N. Buchal and J. B. Keller [10].

The main idea of the boundary-layer method [6] is in stretching the coordinates by some powers (usually fractional) of the large parameter. The derivations are based on the prescribed analytic form of the solution. This representation is substituted into the equation (Helmholtz or Maxwell) and the boundary conditions and leads to a recurrent sequence of simpler problems, which appear by equating terms of similar powers of the large parameter. If the prescribed form is correctly chosen, all these problems can be step-by-step solved up to any desired order. It is important that the solutions should have a sufficient number of arbitrary constants (or functions). This arbitrariness is eliminated when local asymptotic expansions are matched with each other and with the ray representation of the field where it is applicable.

In the case of the homogeneous media, the rays are the straight lines and the main difficulty comes from the surface of the obstacle. Finding the field in such problems of diffraction can be performed in two steps. First, the induced currents on the surface are calculated. Then these currents are substituted into Green's formula (Stratton-Chu formula [28] in electromagnetics) which allows the field to be represented at any point of the free space in the form of the integral over the surface. The first step is the most difficult one and requires the asymptotic technique to be applied, while integration along the surface can be performed even numerically. However, it is possible to apply the saddle point method to such integrals.

In this chapter, we represent classical asymptotic approximations, which are constructed by applying the boundary-layer technique in the form of the parabolic equation method. At the end, we discuss the resulting asymptotic representations having in mind the applicability of them to the problems of diffraction by elongated (or slender) bodies. We discuss the rates of elongation of the bodies and point out the difficulties which the classical parabolic equation method meets when applied to the problems of diffraction by bodies whose elongation rate is sufficiently high.

These difficulties are resolved in the second part of the book, where the ideas of the boundary-layer method are adopted for such problems of diffraction which besides the usual large parameter contains another quantity, which is proportional to some fractional power of the main large parameter.

1.2 Diffraction by a Smooth Convex Body

One of the problems where the boundary-layer method considerations are required is the problem of diffraction by a smooth convex body.

1.2.1 Analysis of the Field of Rays

Let the convex boundary S of a body be illuminated by wave field expressed by its ray expansion

$$u^{\text{inc}} = \frac{A}{\sqrt{J}} \mathrm{e}^{ik\Theta}, \tag{1.1}$$

where J is the divergence of the ray tube and Θ is the eikonal. Figure 1.1 presents the 2D cross-section of the field of rays. At some line on the surface S, the rays of the incident field are tangent to S. We denote the projection of this line on the cross-section as C. The rays which are tangent to the surface are called the limiting rays. The line C separates the surface into illuminated and shadowed parts. The illuminated part is reached by the incident rays and if the incidence is not too close to the tangent, the fields carried by the reflected rays can be described in the frame of the ray method. The shadowed side of the surface is not reached by the incident rays and therefore the ray method approximation predicts there is no field. Thus, the limiting ray outgoing from point C separates the illuminated part of the space and the shadow. In the vicinity of this ray, the ray method fails because it is not possible to distinguish the phases of the incident and the reflected fields. The part of this domain, which is near the surface, is the Fock domain (see Fig. 1.1). In this domain, the parameters of the surface significantly influence the field. The field in the domain which is near to the limiting rays, but sufficiently distant from the surface, is described by the Fresnel asymptotics. The shadowed part of the space is also separated into two parts. Near the surface, the field is described by the creeping wave asymptotic expansion. The rest part of the shadow, i.e. the region between the penumbra and the creeping waves domain, is illuminated by diffracted rays. These rays are launched by creeping

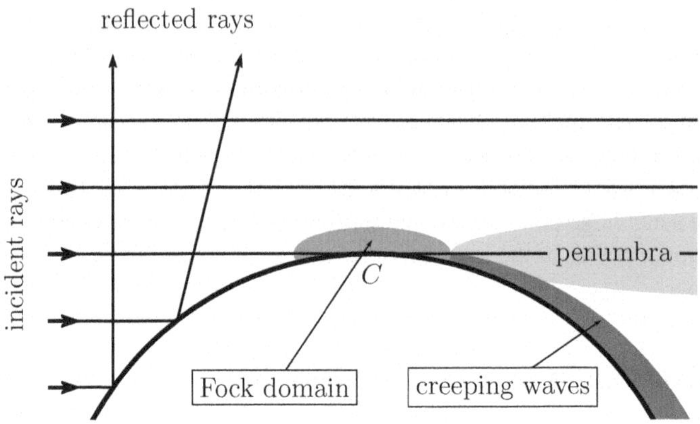

Fig. 1.1 Diffraction by a convex body

waves and follow the tangent lines to the surface. The field of these rays is regular and therefore the ray method is applicable in this region.

Thus, the analysis of the ray field depicted in Fig. 1.1 allows the following domains, where geometrical optics fails, to be noted:

- Penumbra region (Fresnel field);
- Deep shadow near the surface (Creeping waves);
- Small vicinity of curve C—Fock domain.

Local asymptotic expansions of wave fields in all these three domains can be constructed by the boundary-layer method. The field of diffracted rays is expressed by the ray method in which the initial amplitudes should be found by matching with the creeping wave asymptotic representation of the field.

In the following sections, we construct the asymptotics for the field in the Fock domain. The version of the boundary-layer method used in this domain is called the parabolic equation method.

1.2.2 The Parabolic Equation Method

Let us start with the case of acoustic wave diffraction by a cylindrical surface. This reduces the problem to a 2D boundary-value problem for Helmholtz equation

$$\left(\triangle + k^2\right)u = 0.$$

We require some boundary conditions on the surface S. It could be Dirichlet, Neumann or mixed-type condition

$$u\Big|_{n=0}=0,\qquad \frac{\partial u}{\partial n}\Big|_{n=0}=0,\quad \text{or}\quad \frac{\partial u}{\partial n}\Big|_{n=0}+\mathrm{i}kZu\Big|_{n=0}=0.$$

Here, n is the normal coordinate to the surface S and Z in the last variant of boundary condition is the impedance.

To complete the mathematical formulation, some kind of radiation conditions for large n is required. This condition allows only exponentially decreasing or outgoing waves from the surface S.

Now we try to invent the form, in which we search the asymptotic expansion for the solution. This form can be borrowed from the canonical problem of diffraction by a circular cylinder. However in this simple case, we choose it basing on logical considerations. Since both the incident and the reflected rays in the Fock domain go almost along the surface, it is natural to assume that the solution u of the above problem should have the form of the wave process moving along the surface S. We specify this assumption by introducing the representation

$$u=\exp(\mathrm{i}ks)U(s,n), \tag{1.2}$$

where s is the arc-length measured along the surface S, and the new unknown function U called the *attenuation function* is assumed to vary with s coordinate more slowly than the separated exponential multiplier.

In order to substitute representation (1.2) into the Helmholtz equation, the latter should be rewritten in coordinates (s,n)

$$\frac{1}{1+n/\rho}\left(\frac{\partial}{\partial s}\frac{1}{1+n/\rho}\frac{\partial u}{\partial s}+\frac{\partial}{\partial n}\Big(1+n/\rho\Big)\frac{\partial u}{\partial n}\right)+k^2u=0.$$

Here, $\rho=\rho(s)$ is the radius of curvature of surface S. It is convenient to multiply Helmholtz equation by $(1+n/\rho)^2$. Then substituting expression (1.2) yields

$$2\mathrm{i}k\frac{\partial U}{\partial s}+\frac{\partial^2 U}{\partial s^2}+\frac{n}{\rho+n}\frac{\rho'}{\rho}\left(\mathrm{i}kU+\frac{\partial U}{\partial n}\right)+\frac{\partial^2 U}{\partial n^2}+2\frac{n}{\rho}\frac{\partial^2 U}{\partial n^2}+\frac{n^2}{\rho^2}\frac{\partial^2 U}{\partial n^2}$$
$$+\frac{1}{\rho}\frac{\partial U}{\partial n}+\frac{n}{\rho^2}\frac{\partial U}{\partial n}+2k^2\frac{n}{\rho}U+k^2\frac{n^2}{\rho^2}U=0. \tag{1.3}$$

In this cumbersome equation, some terms are more important than the others. If the less important terms are dropped out, the resulting approximate equation will lead to an approximate solution, which can be then corrected by adding smaller order corrections. This means that the solution U can be searched in the form of asymptotic series

$$U=U_0+U_1+U_2+\cdots,$$

where U_1 is in some sense smaller than U_0 and U_2 is smaller than U_1, etc.

Having in mind this approach, it is necessary to decide what terms in (1.3) are the principal ones. To do that, one can exploit different approaches. For example, as we mentioned before it is possible to spy on the structure of the solution in an appropriate canonical problem. However, in this relatively simple case, the significant terms can be picked out by purely deductive analysis. For that, we assume that function U in reality depends on some stretched coordinates $k^\alpha s$ and $k^\beta n$. The powers α and β are positive, but yet unknown. However, since the most quickly varying with s factor is already extracted in (1.2), the power α should be less than one.

Now we consider the terms. As $\alpha < 1$, the term $\partial^2 U/\partial s^2$ is smaller than the term $2ik\partial U/\partial s$ and therefore cannot be considered the principal. Further, with respect to coordinate n, the formulation of the problem has both the boundary condition and the radiation condition. To be able to set two conditions, the partial differential equation should be of the second order with respect to n. That means that at least one term with the second-order derivative $\partial^2 U/\partial n^2$ should be included in the list of the principal order terms. Due to $\beta > 0$, the terms with $\partial U/\partial n$ are smaller than similar terms with $\partial^2 U/\partial n^2$ and terms with n^2 are smaller than similar terms with n in the first power, or those without n. All this allows the following three terms to be considered as pretenders to be the principal terms in the equation

$$\frac{\partial^2 U}{\partial n^2}, \qquad 2ik\frac{\partial U}{\partial s}, \quad \text{and} \quad 2k^2\frac{n}{\rho}U.$$

The first term is necessary in order for the two conditions at $n = 0$ and $n \to +\infty$ to be included in the formulation of the boundary-value problem. The second term is also necessary because if dropping it out we get the ordinary differential equation, which has not enough possibilities to describe the transition of the field from the illuminated to the shadowed part of the Fock domain. And the third term is the only term among the selected pretenders that contains the characteristics of the surface. By dropping it, we would get the solution which does not depend on the geometry of the surface.

Preserving all three terms results in the well-known Leontovich parabolic equation

$$2ik\frac{\partial U}{\partial s} + \frac{\partial^2 U}{\partial n^2} + 2k^2\frac{n}{\rho}U = 0. \tag{1.4}$$

Though this equation is of Shrödinger type, in diffraction theory it is traditionally called the parabolic equation.

1.2.3 Asymptotic Expansion in the Fock Domain

Since all the terms in the parabolic equation (1.4) should have the same order with respect to asymptotic parameter k, this equation fixes the scales $\alpha = 1/3$ and $\beta = 2/3$. So, the Fock domain is as small as $k^{-1/3}$ along the surface and as small as $k^{-2/3}$ in the direction of the normal. For the case of a usual body, one can represent the radius of curvature $\rho(s)$ in such a small domain by the Taylor series

$$\rho(s) = \rho_0 + \rho_0' s + \frac{1}{2}\rho_0'' s^2 + \cdots$$

and replace ρ in the parabolic equation with the zeroth-order term. That is in the Fock domain, the radius of curvature can be considered constant at the leading order approximation. This means that the real surface can be replaced by its simpler representative—by the circular cylinder of radius ρ_0.

We define the boundary-layer coordinates to be

$$\sigma = m_0 \frac{s}{\rho_0}, \qquad \nu = 2m_0^2 \frac{n}{\rho_0}, \qquad m_0 = \left(\frac{k\rho_0}{2}\right)^{1/3}, \tag{1.5}$$

which reduce the parabolic equation to the most simple form

$$\frac{\partial^2 U(\sigma,\nu)}{\partial \nu^2} + \mathrm{i}\frac{\partial U(\sigma,\nu)}{\partial \sigma} + \nu U(\sigma,\nu) = 0.$$

The large parameter k is involved in (1.5) in powers 1/3 and 2/3, therefore it is natural to expect asymptotic sequence U_j, $j = 0, 1, \ldots$ to have orders $O(k^{-j/3})$. That is, the prescribed form for the field (the Ansatz) in the Fock domain is

$$u = \exp(\mathrm{i}ks) \sum_{j=0}^{N} U_j(\sigma,\nu) m_0^{-j}.$$

Inserting this expression into the Helmholtz equation and ranging terms in accordance with the powers of m_0 yields the recurrent system of equations

$$L_0 U_0 = 0, \quad L_0 U_1 + L_1 U_0 = 0, \quad L_0 U_2 + L_1 U_1 + L_2 U_0 = 0, \ldots$$

Here

$$L_0 = \mathrm{i}\frac{\partial}{\partial \sigma} + \frac{\partial^2}{\partial \nu^2} + \nu,$$

$$L_1 = -\nu \rho_0' \sigma,$$

$$L_2 = \frac{1}{4}\frac{\partial^2}{\partial \sigma^2} + \mathrm{i}\nu \frac{\partial}{\partial \sigma} + \frac{1}{2}\frac{\partial}{\partial \nu} + \frac{1}{2}\rho_0 \rho_0'' \nu \sigma^2 - (\rho_0')^2 \nu \sigma^2 - \frac{3}{4}\nu^2$$

and the rest L_j are the differential operators no higher than of second order by σ and ν.

Let U be searched as the sum of the attenuation functions $U^{\mathrm{inc}} + U^{\mathrm{refl}}$, where U^{inc} corresponds to the incident u^{inc} wave and U^{refl} to the reflected waves. Correspondingly, all U_j can be also represented as $U_j^{\mathrm{inc}} + U_j^{\mathrm{refl}}$. Our nearest goal is to find the representation for U_0^{inc} in coordinates (ς, ν). In the small domain, one can decompose the divergence factor J in the ray expansion (1.1) of the incident wave in the Taylor series which makes almost every incident field look like a plane wave. In the case of plane wave incidence

$$u^{\mathrm{inc}} = A\,\mathrm{e}^{ikz},$$

the eikonal $\Theta = z$ can be represented as

$$z = s + \frac{ns}{\rho_0} - \frac{s^3}{6\rho_0^2} + \frac{\rho_0' s^4}{8\rho_0^3} - \frac{\rho_0' n s^2}{2\rho_0^2} + \cdots \tag{1.6}$$

Passing to variables (σ, ν) yields the following Representation:

$$u^{\mathrm{inc}} = A\exp(ikz) \sim A\exp\left(iks + \mathrm{i}\left(\sigma\nu - \sigma^3/3\right)\right).$$

That is

$$U_0^{\mathrm{inc}} = A\exp\left(\mathrm{i}\left(\sigma\nu - \sigma^3/3\right)\right).$$

The leading order term, though computed for the case of plane wave incidence, remains the same for an arbitrary incident wave which has no caustics or focal points in the Fock domain.

Thus at the leading order approximation, we get the following boundary-value problem for the attenuation function U_0^{refl}: it satisfies the parabolic equation

$$L_0 U_0^{\mathrm{refl}} = 0,$$

the radiation condition at infinity $\nu \to \infty$ and the inhomogeneous boundary condition (either Dirichlet, Neumann or impedance)

$$U_0^{\mathrm{refl}}(\sigma, 0) = -A\mathrm{e}^{-\mathrm{i}\sigma^3/2}, \qquad \frac{\partial U_0^{\mathrm{refl}}(\sigma, 0)}{\partial \nu} = -\mathrm{i}A\sigma\mathrm{e}^{-\mathrm{i}\sigma^3/2}$$

or

$$\frac{\partial U_0^{\mathrm{refl}}(\sigma, 0)}{\partial \nu} + \mathrm{i}q U_0^{\mathrm{refl}}(\sigma, 0) = -\mathrm{i}A(\sigma + q_0)\mathrm{e}^{-\mathrm{i}\sigma^3/2},$$

where

$$q_0 = m_0 Z.$$

In the latter case, we consider the impedance Z to be such that $q_0 = O(1)$.

Applying Fourier transform with respect to σ to this boundary-value problem reduces it to the problem for the ordinary differential equation

$$\frac{\partial^2 \hat{U}_0}{\partial \nu^2} + (\nu - \xi)\hat{U}_0 = 0.$$

This is Airy equation [1], and its standard solutions are $\mathrm{Ai}(\xi-\nu)$ and $\mathrm{Bi}(\xi-\nu)$. However, it is more convenient to use Airy functions in Fock notations (see Appendix A.1). Then, in view of the integral representation (A.3) for the Airy function v, we get

$$U_0^{\mathrm{inc}} = \frac{A}{\sqrt{\pi}} \int_{-\infty}^{+\infty} \mathrm{e}^{\mathrm{i}\xi\sigma} v(\xi-\nu)\, \mathrm{d}\xi.$$

The reflected field is searched in the form

$$U_0^{\mathrm{refl}} = \int_{-\infty}^{+\infty} \mathrm{e}^{\mathrm{i}\zeta\varsigma} A_0(\xi) w_1(\xi-\nu)\, \mathrm{d}\xi.$$

The choice of Airy function w_1 in that formula is due to the radiation condition for $\nu \to +\infty$ and the asymptotic expansion (A.5)

$$w_1(-z) \sim z^{-1/4} \exp\left(\frac{2\mathrm{i}}{3} z^{3/2} + \frac{\pi \mathrm{i}}{4}\right).$$

Function $A_0(\xi)$ is defined from the boundary condition on the surface. For the surface described by the Dirichlet boundary condition, it is

$$A_0(\xi) = -\frac{A}{\sqrt{\pi}} \frac{v(\xi)}{w_1(\xi)}.$$

In the case of the Neumann boundary condition, it is

$$A_0(\xi) = -\frac{A}{\sqrt{\pi}} \frac{\dot{v}(\xi)}{\dot{w}_1(\xi)}.$$

In the case of the impedance boundary condition with $Z = O(k^{-1/3})$, it is

$$A_0(\xi) = -\frac{A}{\sqrt{\pi}} \frac{\dot{v}(\xi) - \mathrm{i}q_0\, v(\xi)}{\dot{w}_1(\xi) - \mathrm{i}q_0\, w_1(\xi)},$$

where $q_0 = m_0 Z$.

Combining the above formulas yields the following expression for the field in the principal order by k

$$u_0 = \frac{A}{\sqrt{\pi}} \mathrm{e}^{iks} \int_{-\infty}^{+\infty} \mathrm{e}^{\mathrm{i}\xi\sigma} \left(v(\xi-\nu) - \frac{v(\xi)}{w_1(\xi)} w_1(\xi-\nu) \right) \mathrm{d}\xi \tag{1.7}$$

in the case of acoustically soft surface,

$$u_0 = \frac{A}{\sqrt{\pi}} e^{iks} \int_{-\infty}^{+\infty} e^{i\xi\sigma} \left(v(\xi - \nu) - \frac{\dot{v}(\xi)}{\dot{w}_1(\xi)} w_1(\xi - \nu) \right) d\zeta$$

in the case of acoustically hard surface and

$$u_0 = \frac{A}{\sqrt{\pi}} e^{iks} \int_{-\infty}^{+\infty} e^{i\xi\sigma} \left(v(\xi - \nu) - \frac{\dot{v}(\xi) - iq_0\, v(\xi)}{\dot{w}_1(\xi) - iq_0\, w_1(\xi)} w_1(\xi - \nu) \right) d\xi \qquad (1.8)$$

In the case of the impedance boundary condition. Below, we consider only the case of the impedance boundary condition. Then by setting $q_0 = 0$ and $q_0 \to \infty$, one can get correspondingly the formulas for the cases of the Neumann and the Dirichlet boundary conditions.

We note also that in the case of large positive σ, the above integrals can be computed with the use of the residue theorem, which requires the study of the zeros of Airy function w_1, its derivative $\dot{w}_1$ and solutions of the dispersion equation

$$\dot{w}_1(\xi) - iq_0 w_1(\xi) = 0$$

(see Appendix A.1).

Formulas (1.7)–(1.8) are formal, in particular for the next order terms of the asymptotic series the integral should be understood in a special way because the integrand increases when $\xi \to -\infty$ [6]. The regularization of the above integrals is performed as follows. For negative ξ, the Airy function $v(\xi)$ in the expression for the reflected field is expressed via Airy functions w_1 and w_2 as $(w_1(\xi) - w_2(\xi))/(2i)$. Then the integral is split into three

$$\begin{aligned} u_0^{\text{refl}} &= \frac{Ai}{2\sqrt{\pi}} \int_{-\infty}^{0} e^{i\xi\sigma} w_1(\xi - \nu)\, d\xi \\ &\quad - \frac{Ai}{2\sqrt{\pi}} \int_{-\infty}^{0} e^{i\xi\sigma} \frac{\dot{w}_2(\xi) - iq_0 w_2(\xi)}{\dot{w}_1(\xi) - iq_0\, w_1(\xi)} w_1(\xi - \nu)\, d\xi \\ &\quad - \frac{A}{\sqrt{\pi}} \int_{0}^{+\infty} e^{i\xi\sigma} \frac{\dot{v}(\xi) - iq_0 v(\xi)}{\dot{w}_1(\xi) - iq_0\, w_1(\xi)} w_1(\xi - \nu)\, d\xi. \end{aligned}$$

The path of integration in the first integral is shifted to the ray $\ell_1 = (\infty e^{-2\pi i/3}, 0]$, and the path of integration in the second integral is shifted to the ray $\ell_2 = (\infty e^{2\pi i/3}, 0]$ where integrands exponentially decrease.

After constructing the asymptotic representations for the fields, it is a simple matter to find the induced currents

$$u|_{n=0} = \frac{A}{\sqrt{\pi}} e^{iks} \int_{-\infty}^{+\infty} \frac{e^{i\xi\sigma}}{\dot{w}_1(\xi) - iq_0 w_1(\xi)} d\xi, \tag{1.9}$$

$$\left.\frac{\partial u}{\partial n}\right|_{n=0} = -\frac{2iq_0 m_0^2}{\rho_0} \frac{A}{\sqrt{\pi}} e^{iks} \int_{-\infty}^{+\infty} \frac{e^{i\xi\sigma}}{\dot{w}_1(\xi) - iq_0 w_1(\xi)} d\xi. \tag{1.10}$$

The derivations of these formulas took into account the value of the Wronskian (A.6)

$$v(\xi)\dot{w}_1(\xi) - \dot{v}(\xi)w_1(\xi) = 1.$$

By setting $q_0 = 0$ and $q_0 = \infty$, we identify the classical Fock integrals

$$f(\sigma) = \frac{1}{\sqrt{\pi}} \int_{-\infty}^{+\infty} \frac{e^{i\xi\sigma}}{w_1(\xi)} d\xi, \qquad g(\sigma) = \frac{1}{\sqrt{\pi}} \int_{-\infty}^{+\infty} \frac{e^{i\xi\sigma}}{\dot{w}_1(\xi)} d\xi$$

tabulated in [15].

1.2.4 Generalizations to 3D and Electromagnetic Problems

In the 3D case, we have a set of limiting rays. Each of these rays defines a geodesic on the surface, which crosses the light-shadow boundary C and goes in the same direction as the limiting ray. This set of geodesics is usually regular in the vicinity of the curve C. Let the coordinate system (s, α) on the surface be such that coordinate α specifies the geodesics and the coordinate s is the arc-length measured along this geodesic from some reference line. Any point in the exterior of the body is then characterized by coordinates (α, s, n), where, as previously, n is the normal to the surface. The matrix of the second quadratic form of the system (s, α, n) takes the form

$$g_{ij} = \begin{pmatrix} \left(1 + \frac{n}{\rho}\right)^2 + \tau^2 n^2 & -h\tau\left(2n + \frac{n^2}{\rho} + cn^2\right) & 0 \\ -h\tau\left(2n + \frac{n^2}{\rho} + cn^2\right) & h^2\left((1+cn)^2 + \tau^2 n^2\right) & 0 \\ 0 & 0 & 1 \end{pmatrix} + O(n^2),$$

where $\rho = \rho(s, \alpha)$ and $\tau = \tau(s, \alpha)$ are the radius of curvature of the geodesics $\alpha = const$ and its torsion, $c = c(s, \alpha)$, is the normal curvature of the surface in the direction orthogonal to the geodesics (transverse curvature), the parameter

$h = h(s, \alpha)$ measures the narrowing (or broadening) of an infinitesimal geodesics pencil between a point with coordinates (s, α) and the reference curve $s = 0$.

Helmholtz equation in coordinates (s, α, n) takes the form

$$\frac{1}{\sqrt{g}}\frac{\partial}{\partial s}\left(\sqrt{g}g_{ss}\frac{\partial u}{\partial s} + \sqrt{g}g_{s\alpha}\frac{\partial u}{\partial \alpha}\right) + \frac{1}{\sqrt{g}}\frac{\partial}{\partial \alpha}\left(\sqrt{g}g_{\alpha s}\frac{\partial u}{\partial s} + \sqrt{g}g_{\alpha\alpha}\frac{\partial u}{\partial \alpha}\right)$$
$$+ \frac{1}{\sqrt{g}}\frac{\partial}{\partial n}\left(\sqrt{g}\frac{\partial u}{\partial n}\right) + k^2 u = 0, \tag{1.11}$$

where $g = \det(g_{ij})$.

The solution is searched in the form similar to (1.2), namely

$$u = \mathrm{e}^{\mathrm{i}ks}U(\sigma, \alpha, \nu) \tag{1.12}$$

and the attenuation function U is assumed to depend on stretched coordinates (1.5). However, now ρ_0 and hence m_0 depend on α.

We decompose other characteristics of the surface into the Taylor series with respect to s

$$h = h_0(\alpha) + h_0'(\alpha)s + \cdots, \qquad c = c_0(\alpha) + \cdots, \qquad \tau = \tau_0(\alpha) + \cdots.$$

Inserting representation (1.12) into Helmholtz equation and collecting terms in accordance with their order, one gets

$$\left(L_0 + m_0^{-1}L_1 + m_0^{-2}L_2 + \cdots\right)U = 0.$$

Here

$$L_0 = \mathrm{i}\frac{\partial}{\partial \sigma} + \frac{\partial^2}{\partial \nu^2} + \nu,$$

$$L_1 = -\nu\rho_0'\sigma + \frac{\mathrm{i}\rho_0}{2}\frac{h_0'}{h_0}$$

and L_2 is a cumbersome differential operator that besides ρ_0, ρ_0' and ρ_0'', as in 2D case, includes such characteristics of the surface as transverse curvature c_0, torsion τ_0 and the derivative of the spreading factor h_0'.

Since the leading order operator L_0 appears the same as in the 2D case, the leading order approximation is given by exactly the same formulas as in the 2D case, but the transverse coordinate α is involved in these expressions as a parameter, on which the amplitude A, the radius of curvature ρ_0 and possibly the impedance Z depend.

In the case of electromagnetic wave diffraction, it is necessary to rewrite Maxwell equations

$$\begin{cases} \nabla \times \mathbf{E} = \mathrm{i}k\mathbf{H}, \\ \nabla \times \mathbf{H} = -\mathrm{i}k\mathbf{E}, \end{cases}$$

in coordinates (s, α, n). The rotation operator $\nabla\times$ has the form

$$(\nabla \times \mathbf{E})_s = \frac{g_{ss}}{\sqrt{g}}\left(\frac{\partial E_n}{\partial \alpha} - \frac{\partial E_\alpha}{\partial n}\right) + \frac{g_{s\alpha}}{\sqrt{g}}\left(\frac{\partial E_s}{\partial n} - \frac{\partial E_n}{\partial s}\right),$$

$$(\nabla \times \mathbf{E})_\alpha = \frac{g_{\alpha s}}{\sqrt{g}}\left(\frac{\partial E_n}{\partial \alpha} - \frac{\partial E_\alpha}{\partial n}\right) + \frac{g_{\alpha\alpha}}{\sqrt{g}}\left(\frac{\partial E_s}{\partial n} - \frac{\partial E_n}{\partial s}\right),$$

$$(\nabla \times \mathbf{E})_n = \frac{1}{\sqrt{g}}\left(\frac{\partial E_\alpha}{\partial s} - \frac{\partial E_s}{\partial \alpha}\right).$$

The components E_s, E_n, H_s and H_n of the electric and magnetic vectors can be expressed via the transverse components E_α and H_α, for which we again obtain at the leading order the parabolic differential equation with operator L_0.

1.2.5 The Higher Order Corrections

In order to find the next order corrections, it is necessary first to obtain the next order terms in the representation of the incident wave. These terms can be found starting from the representation (1.6) for the eikonal of the incident wave in the boundary-layer coordinates. For the case of the plane wave incidence,

$$U_1^{\text{inc}} = U_0^{\text{inc}} \frac{\mathrm{i}\rho_0'}{4}\left(\sigma^4 - 2\sigma^2\nu\right).$$

After computing the inverse Fourier transform, one obtains the representation for U_1^{inc}, which can be written in the following form:

$$U_1^{\text{inc}} = \frac{A}{\sqrt{\pi}}\frac{\mathrm{i}\rho_0'}{4}\int_{-\infty}^{+\infty} \mathrm{e}^{\mathrm{i}\xi\sigma}\left[(\xi^2 - \nu^2)v(\xi - \nu) + 2\dot{v}(\xi - \nu)\right] \mathrm{d}\xi.$$

In the case of the arbitrary incident wave, the formula becomes more complicated and involves the divergence factor J.

The second and higher order terms in the asymptotic representation of the incident field u^{inc} have similar structure

$$U_j^{\text{inc}} = \frac{A}{\sqrt{\pi}}\int_{-\infty}^{+\infty} \mathrm{e}^{\mathrm{i}\xi\sigma}\left[P_j^{\text{inc}}(\xi, \nu)v(\xi - \nu) + Q_j^{\text{inc}}(\xi, \nu)\dot{v}(\xi - \nu)\right] \mathrm{d}\xi,$$

where P_j^{inc} and Q_j^{inc} are polynomials with respect to ξ and ν. One can check that at the order $j = 2$, polynomials P_j^{inc} and Q_j^{inc} have the orders of 4 and 2. See [21], where the simpler case of plane wave incidence on the surface of a body of revolution is considered.

The next order corrections U_1^{refl}, U_2^{refl}, …for the reflected field satisfy inhomogeneous parabolic equations

$$L_0 U_j = f_j,$$

with the right-hand sides f_j which depend on the leading order term

$$f_1 = -L_1 U_0, \qquad f_2 = -L_2 U_0 - L_1 U_1, \qquad \dots$$

Applying Fourier transform with respect to σ reduces these equations to the inhomogeneous Airy equation

$$w''(\nu) + (\nu - \xi) w(\nu) = \hat{f}_j$$

with the right-hand sides containing Airy functions and their derivatives. Therefore, U_j are given by Fourier integrals

$$U_j^{\mathrm{refl}} = \frac{1}{\sqrt{\pi}} \int_{-\infty}^{+\infty} \mathrm{e}^{\mathrm{i}\xi\sigma} \left[\left(P_j(\xi, \nu) + A_1(\xi) \right) w_1(\xi - \nu) + Q_j(\xi, \nu) \dot{w}_1(\xi - \nu) \right] \mathrm{d}\xi$$

containing polynomials P_j and Q_j of ν and ξ.

For the first-order correction U_1^{refl}, the right-hand side is

$$\hat{f}_1 = -\mathrm{i}\rho_0' \nu \frac{\partial A_0(\xi) w_1(\xi - \nu)}{\partial \xi} - \frac{\mathrm{i}\rho_0}{2} \frac{h_0'}{h_0} A_0(\xi) w_1(\xi - \nu).$$

Using Table A.7, one finds

$$P_1(\xi, \nu) = \frac{\mathrm{i}}{2} \rho_0' A_0(\xi) \nu^2,$$

$$Q_1(\xi, \nu) = -\frac{\mathrm{i}}{3} \rho_0' \dot{A}_0(\xi) (4\xi - \nu) - \frac{\mathrm{i}}{2} A_0(\xi) \left(\rho_0' + \rho_0 \frac{h_0'}{h_0} \right).$$

Here $\dot{A}_0(\xi) = \mathrm{d}A_0(\xi)/\mathrm{d}\xi$.

The arbitrariness of A_1 is eliminated when the boundary conditions are satisfied. For example, in the case of the Dirichlet boundary condition,

$$A_1(\xi) = -P_1(\xi, 0) - P_1^{\mathrm{inc}}(\xi, 0) \frac{v(\xi)}{w_1(\xi)} - Q_1(\xi, 0) \frac{\dot{w}_1(\xi)}{w_1(\xi)} - Q_1^{\mathrm{inc}}(\xi, 0) \frac{\dot{v}_1(\xi)}{w_1(\xi)}.$$

Already at the order $j = 1$ the formulas become very cumbersome. At the next orders, the complexity dramatically increases. Several attempts are known to derive the corrections to the Fock asymptotic expansion. The first was undertaken in 1967 by S. Hong [16]. He was interested in computing the current on the surface and used a different approach based on the integral equations for the currents in which he assumed that the geodesics of the surface have no torsion and $\rho_0' = 0$. He was able to derive the second-order terms of the asymptotic expansions for the fields in the problems of acoustic diffraction by a hard surface and in the case of electromagnetic wave diffraction by a perfect conductor. Unfortunately, his formulas contain some misprints. Analysis of the second-order corrections in a more general case of electromagnetic creeping waves on an impedance surface was undertaken in [3], and the results are included in the book [26]. The most recent results for the second-order corrections in the problem of diffraction by a soft body of revolution can be found in [21]. An attempt to write the whole series of corrections for the case of diffraction by the circular cylinder and the sphere was undertaken in [18], but the formulas of this paper disagree with those of [16].

We do not present here the corrections to Fock asymptotics, however, it is important to summarize the structure of these terms. As we can already confirm from the derivations presented above, the leading order approximation appears the same in the case of the 2D diffraction, the 3D diffraction of acoustic wave and the 3D diffraction of electromagnetic waves. It depends only on the radius of curvature of the geodesics at the light-shadow boundary. The first-order corrections involve such characteristics as the variation of curvature, expressed by ρ_0', the divergence of the geodesics h and the divergence of the incident ray tube J. Besides, in this order the impurity component of the electromagnetic field manifests itself. Such characteristics as transverse curvature c of the surface and torsion τ of the geodesics on the surface appear only at the second-order correction U_2. Therefore, in an ordinary case, the influence of the transverse curvature and the torsion on the diffracted field is as small as $(k\rho_0)^{-2/3}$.

1.3 Accuracy of the Fock Approximation

Usually, there is no need to use higher order corrections to Fock asymptotics because its leading order term itself provides a sufficiently accurate approximation. As was explained in [17] for the problem of diffraction by a circular cylinder, Fock approximation is applicable in the case of $k\rho \approx 3$ and higher. This is a bit surprising observation because the asymptotic parameter m_0 in the case of $k\rho = 3$ is only 15% larger than unity. Nevertheless, figure 111d from [17], which we copy here as Fig. 1.2, confirms that the relative error of the leading order term approximation for the induced currents on the perfectly conducting cylinder in the case of the incidence of a plane TM electromagnetic wave is less than 5%.

This can be explained by noting that in this particular problem, the quantity ρ_0' is equal to zero and the first-order correction U_1 vanishes. However, in the case of diffraction by a sphere when ρ_0' is also equal to zero, the lower bound of parameter

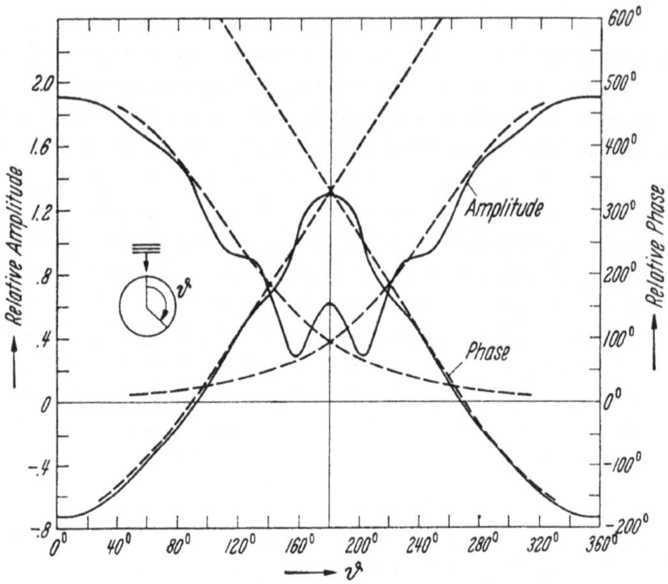

Fig. 1.2 Comparison of Fock asymptotics with the exact values of the currents on a perfectly conducting cylinder with $k\rho = 3.1$ in TM case. Reprinted with permission from [17]. Copyright ©1987, Springer-Verlag OHG. Berlin-Gottingen-Heidelberg

$k\rho$ starting from which the leading order term provides similar good approximation shifts to the values of 10–15 [8]. In the case of diffraction by prolate spheroids, this bound further increases. We illustrate this observation in Fig. 1.3, which shows the induced currents for the case of axial incidence on perfectly conducting spheroids with different aspect ratios.

On the top, the spheroid is with the semiaxes of 2.5 m and 1 m and the frequency is 1 GHz. So, $\rho_0 = 6.25$ m, $k\rho_0 \approx 130$ and Fock parameter $m_0 \approx 4$. On the right, the semiaxes are 1.77 m and 0.5 m, while the frequency is 2 GHz. So, ρ_0 is the same as for the first spheroid, $k\rho_0 \approx 260$ and Fock parameter $m_0 \approx 5$. We see that in both cases Fock approximation significantly underestimates the amplitudes of the currents. Moreover, the errors appear larger in the case of the second spheroid than in the case of the first one, while the asymptotic parameter m_0 is larger in the second case.

Analogous comparison of the leading order term asymptotics with the fields of diffraction by acoustically soft and hard spheroids shows that in the case of soft surface Fock approximation overestimates real fields and in the case of hard surface it gives noticeably lower values.

Certainly, by increasing the frequency in the problem of diffraction by any prolate spheroid with an arbitrarily large aspect ratio, we come to the situation where Fock asymptotics is accurate. However, the required frequency may be too high. On this hand, it is important to derive an approximation for the induced currents on an

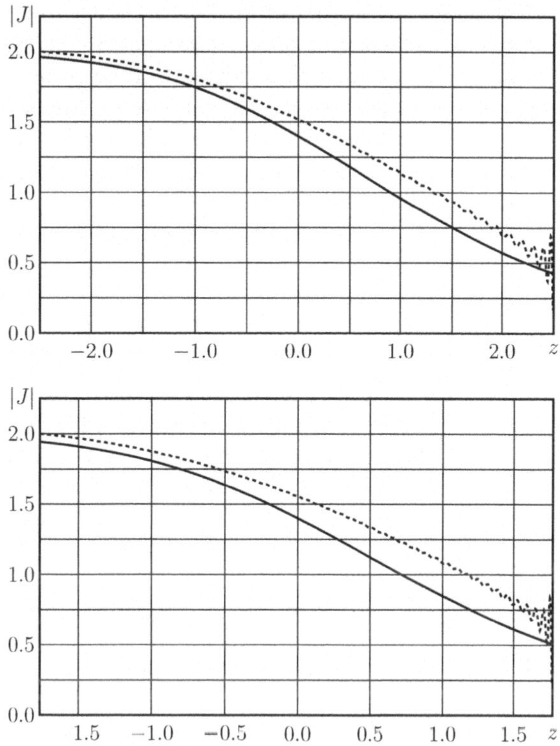

Fig. 1.3 Fock approximation and numerically computed currents on perfectly conducting spheroids at axial incidence. On the top: $b = 2.5$ m, $a = 1$ m and frequency is 1 GHz; on the bottom $b = 1.77$ m, $a = 0.5$ m and frequency is 2 GHz

elongated body that would be applicable at not that high frequencies. Problems of that type have been also called the problems of diffraction by slender bodies [12].

1.4 The Transverse Curvature Effects

1.4.1 Transverse Curvature in the Classical Fock Asymptotic Approximation

The results of the previous section show that the rate of elongation affects the accuracy of the leading order term of the Fock asymptotic expansion. Achieving similar accuracy for a more elongated spheroid requires the frequency to be higher. For very much elongated spheroids with aspect ratios of 10:1 and higher, the results predicted by the Fock approximation may be absolutely wrong even in the case of sufficiently large values of $k\rho_0$.

Comparing the quality of Fock asymptotics to approximate field on a circular cylinder, sphere, and less and more elongated spheroids, one can naturally conclude that the main characteristics which affect the accuracy of the approximation is the transverse curvature c of the surface. For the case of the cylinder it is zero, on the sphere it is $c = 1/\rho_0$ and for the case of the much elongated spheroid it can be arbitrarily large. As we noted above, the transverse curvature c is presented in the asymptotic expansion only starting from the second-order term U_2. For the usual body, like a sphere, its contribution is proportional therefore to $c(k\rho_0)^{-2/3}$. To form the parameter responsible for the effect of transverse curvature independent of the unit length, we define it as

$$C = m_0^{-2}\rho_0 c_0.$$

For the usual body, C is small and to analyze the effect of transverse curvature we need to study the second-order terms U_2. If c_0 increases up to the order of m_0, parameter C becomes of order $O(m_0^{-1})$ and the corresponding terms move to the first- order correction U_1. This case in [3] was called to be the case of diffraction by a *moderately elongated body*. The other critical change happens if $C = O(1)$. This case in [3] was called to be the case of diffraction by a *strongly elongated body*.

1.4.2 Moderately Elongated Bodies

The parabolic equation method applied to a problem of diffraction by a moderately elongated body requires no modification because the principal order terms of the differential equation remain the same. However, compared to the classical case operator L_1 changes by including the following term:

$$L_1^c = \frac{c_0\rho_0}{2m_0}\frac{\partial}{\partial \nu}.$$

This term is originated from the approximation $\sqrt{g} \approx h(1 + cn)$ and appears when in the term containing derivatives with respect to n in (1.11) the outer differentiation is performed on $\sqrt{g}$. The right-hand side f_1 of the inhomogeneous differential equation for U_1 gets the addition

$$f_1^c = \frac{c_0\rho_0}{2m_0} A_0(\xi)\dot{w}_1(\xi - \nu).$$

To analyze its influence on the solution U_1, let us consider the case of the impedance boundary condition, which is in some sense uniform and enables the cases of the Dirichlet or the Neumann boundary conditions to be obtained by taking the limits as $Z \to \infty$ and $Z \to 0$ in the final formulas for the currents. The term f_1^c in the right-hand side of the equation

$$L_0 U_1 = f_1$$

in the case of the impedance surface adds to U_1^{refl} the following function:

$$U_1^c = \frac{c_0 \rho_0}{4m_0\sqrt{\pi}} \int_{-\infty}^{+\infty} \mathrm{e}^{\mathrm{i}\xi\sigma} A_0(\xi) \left[\frac{w_1(\xi)}{\dot{w}_1(\xi) - \mathrm{i}q_0 w_1(\xi)} + \nu \right] w_1(\xi - \nu)\mathrm{d}\xi.$$

This addition agrees with the results obtained by S. Hong in [16].

In the asymptotic approximation for the induced currents, the most simple way to take into account this addition is to introduce in (1.9) and (1.10) the effective impedance

$$Z_{\text{eff}} = Z + \frac{\mathrm{i}}{4} \frac{c_0 \rho_0}{m_0^3} = Z + \frac{\mathrm{i}}{2} \frac{c_0}{k}. \tag{1.13}$$

In the case of electromagnetic wave diffraction, one can find the term that expresses the effect of transverse curvature in a similar way. It results in the same addition to the field, but with the opposite sign. Thus, the effective impedance for the electromagnetic problem is

$$Z_{\text{eff}} = Z - \frac{\mathrm{i}}{2} \frac{c_0}{k}. \tag{1.14}$$

The effective impedances (1.13) and (1.14) coincide with those found in [4] basing on the analysis of the creeping wave asymptotics which describe the field in the deep shadow near the surface of a moderately elongated body (see also [26]).

It is important to note that the effect of the transverse curvature is smaller by one order of $k^{1/3}$ in the case of acoustic wave diffraction by a soft surface, the currents on which are obtained by taking the limit $Z \to \infty$ in (1.8), because in that case the addition to the effective impedance (1.13) becomes negligible. The same is true in the case of the TE polarized electromagnetic wave diffraction by a perfect electric conductor.

Introducing the effective impedance to the Fock asymptotic expansion improves the accuracy of the approximation. For the case of diffraction of the TM polarized electromagnetic wave, this can be seen from the results, presented in Fig. 1.4. However, a noticeable error remains. For more elongated spheroids, the error of the Fock approximation is modified by introducing the effective impedance increases, which shows that a more accurate approximation for the induced currents is necessary.

1.4.3 Strongly Elongated Bodies

The case of a strongly elongated body, characterized by $C = O(1)$, crucially changes the principal order terms in Helmholtz or Maxwell equations. Transverse curvature manifests itself already at the leading order operator, which if written in the usual boundary-layer coordinates (1.5) contains singular terms. These terms make variable separation to be impossible in the general case. The structure of geodesics on the

Fig. 1.4 Fock approximation with the effective impedance $Z_{\text{eff}} = -1$ and the numerically computed currents on the same spheroids as in Fig. 1.3

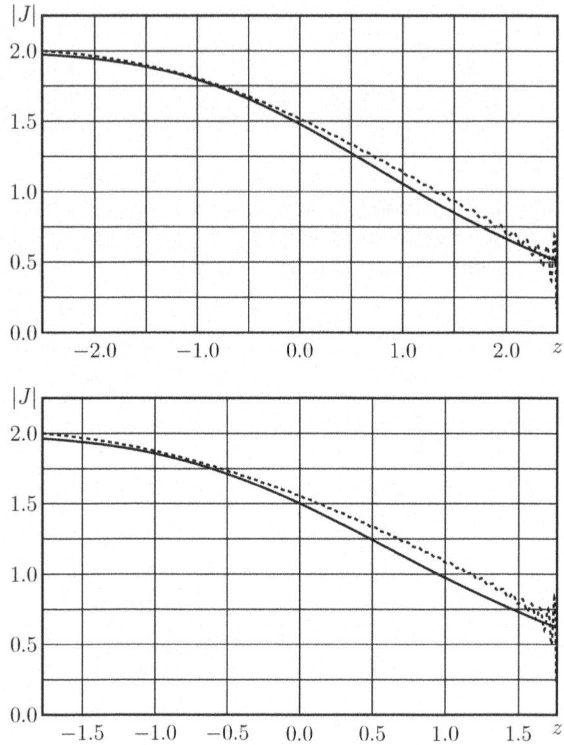

surface may become complicated. For axial incidence on a body of revolution, the geodesics are the "meridians" and have focal points at the tips. For skew incidence, these focal points are deformed into caustics with cusp points. Therefore, the angle of incidence with respect to the direction of elongation becomes an important parameter of the problem. To understand better the difficulties and the physical effects that may be expected in the problems of diffraction by elongated bodies, it is convenient to consider a simple representative surface. In the same manner as the circular cylinder well represents the surface of a usual convex body in the Fock domain, let the spheroid with the appropriately chosen semiaxes represent the elongated body.

Consider first the case of axial incidence. To have the same value of the curvature, we need to impose the relation

$$\frac{b^2}{a} = \rho_0$$

to the semiaxes: major b and minor a of the spheroid. The smaller semiaxis evidently coincides with $1/c_0$. Thus, we define the semiaxes of the representative spheroid as

$$b = \sqrt{\frac{\rho_0}{c_0}}, \qquad a = \frac{1}{c_0}.$$

The wave size of the body is then characterized by the parameters

$$kb = \frac{2}{\sqrt{C}} m_0^2, \qquad ka = \frac{2}{C} m_0. \tag{1.15}$$

This allows making the important conclusion: In the case of diffraction by a strongly elongated body, the whole surface lies inside the Fock domain. In particular, it means that such bodies do not produce deep shadow, but only the penumbra and as a sequence, generally speaking, it is impossible to identify creeping waves on such surfaces. There is only a transition zone to creeping wave asymptotics in the rear part of the Fock domain.

In the classical problems of diffraction, the characteristic wave size of the body is chosen as $k\rho$. As we see from (1.15), this size much overestimates the real wave size of the elongated spheroid. Therefore, it is more convenient to characterize the problems of diffraction by strongly elongated bodies with the asymptotic parameter kb. Then, instead of the characteristic parameter C, it is more convenient to introduce parameter χ, which depends on the rate of elongation, as

$$\widetilde{\chi} = \frac{(ka)^2}{kb} = 2C^{-3/2}.$$

We traditionally refer to $\widetilde{\chi}$ as to the *elongation parameter*, though in view that the more elongated bodies have smaller χ, it is more natural to call it the parameter of "leanness" or "slenderness".

The distance between the focuses of a strongly elongated spheroid coincides with its length up to an asymptotically small correction. Therefore, one can use kp, where $p = \sqrt{b^2 - a^2}$ is the half of the inter-focal distance of the representative spheroid, and define the elongation parameter as

$$\chi = \frac{(ka)^2}{kp}.$$

As we shall see from the following chapters, this allows the asymptotic formulas to be written in a bit more compact form. The parameters $\widetilde{\chi}$ and χ almost coincide and sometimes we do not make the difference between them.

As was already noted, the direction of incidence with respect to the direction of elongation is another important characteristics of the problems of diffraction by strongly elongated bodies. Evidently, the effects of the large transverse curvature manifest themselves only in the paraxial regime. Therefore, we require that the main direction of wave propagation is at a small angle ϑ_0 to the major axis of the representative spheroid. We require this smallness to be characterized by the condition

$$|\vartheta_0|\sqrt{kp} < const, \tag{1.16}$$

which means that in the case of higher frequencies we can consider only such waves that run at a smaller angle to the axis of the body.

One can generalize the definition of the strongly elongated body to such bodies whose transverse wave size has the order of the square root of the longitudinal one. For such bodies, the effects related to large transverse curvature c generally speaking cannot be separated from those associated with the divergence h' and variation of ρ. So, all these effects and the corresponding "correction" terms in the Fock asymptotic expansion should be studied together.

1.5 Conclusion

In this chapter, we have presented the derivation of the leading order term in the classical Fock asymptotic expansion for the currents induced by a high-frequency wave in the vicinity of the light-shadow boundary on the surface of a convex body. We have discussed the correction terms to this asymptotic approximation and have identified the parameters of the geometry which may affect the amplitudes of the induced currents. Since our particular interest is in the analysis of the effects related to large transverse curvature, we distinguish three cases: the case of an ordinary body, the case of a moderately elongated body and the case of a strongly elongated body. In the case of a moderately elongated body, the classical asymptotic procedure remains applicable and results in the approximation, which introduces the effective impedance, which reproduces the first-order terms associated with the transverse curvature c, i.e. the terms that would be of the order of $O(m_0^{-1})$ in the ordinary case. For the case of a strongly elongated body, we see that the usual asymptotic procedure faces serious difficulties. However, these difficulties will be overcome in the case of some particular geometries in the following chapters of this book, where we consider either 2D problems (see Chap. 2) or the case of strongly elongated bodies of Revolution. However, the technique developed there can be generalization to the case of an elongated 3-axis ellipsoid [2], but in this case the Whittaker functions are replaced with the more complicated functions which are the solutions of the confluent Heun equation.

To conclude his chapter, we note that asking C to be larger than in the case of a strongly elongated body by one more power of parameter m_0 results in such geometries when the transverse cross-section of the body is comparable to the wavelength. This resonant regime can be treated only numerically.

References

1. M. Abramowitz, I.A. Stegun (ed.), *Handbook of Mathematical Functions with Formulas, Graphs and Mathematical Tables*, National Bureau of Standards, Applied Mathematics Series 55 (1964)
2. I.V. Andronov, N.I. Andronov, Plane wave diffraction by a strongly elongated three-axis ellipsoid. Acoust. Phys. **67**(4), 341–350 (2021)

3. I.V. Andronov, D. Bouche, Asymptotic of creeping waves on a strongly prolate body. Ann. Télécommun. **49**, 205–210 (1994)
4. I.V. Andronov, D. Bouche, Creeping waves on strongly prolate bodies, in *Proceedings of the URSI EMTS* (St. Petersburg, Russia, 1995), pp. 404–406
5. V.M. Babich, V.S. Buldyrev, *Short-Wavelength Diffraction Theory: Asymptotic Methods* (Springer, Berlin, 1991)
6. V.M. Babich, N.Ya. Kirpichnikova, *The Boundary-Layer Method in Diffraction Problems* (Springer, Berlin, 1979)
7. V.M. Babich, M.A. Lyalinov, V.E. Grikurov, *Diffraction Theory: The Sommerfeld-Malyuzhinets Technique* (Alpha Science, Oxford, UK, 2008)
8. M.G. Belkina, Radiation characteristics of an elongated rotary ellipsoid, in *Diffraction of Electromagnetic Waves on Certain Bodies of Revolution* (Soviet Radio, Moscow, 1957)
9. H. Bremmer, S.W. Lee, Geometrical optics solution of reflection from arbitrarily curved surface. Radio Sci. **17**(5), 1117–1131 (1982)
10. R.N. Buchal, J.B. Keller, Boundary layer problems in diffraction theory. Commun. Pure Appl. Math. **13**, 85 (1960)
11. G.A. Dechamps, Ray techniques in electromagnetics. Proc. IEEE **60**(9), 1022–1035 (1972)
12. J.C. Engineer, J.R. King, R.H. Tew, Diffraction by slender bodies. Eur. J. Appl. Math. **9**, 129–158 (1998)
13. V.A. Fock, The distribution of currents induced by a pane wave on the surface of a conductor. J. Phys. USSR **10**(2), 130–136 (1946)
14. V.A. Fock, New methods in diffraction theory. Philos. Mag. Ser. **7**(39), 149 (1948)
15. V.A. Fock, *Electromagnetic Diffraction and Propagation Problems* (Pergamon Press, Frankfurt, 1965)
16. S. Hong, Asymptotic theory of electromagnetic and acoustic diffraction by smooth convex surfaces of variable curvature. J. Math. Phys. **8**(6), 1223–1232 (1967)
17. H. Hönl, A.W. Maue, K. Westpfahl, *Theorie der Beugung* (Springer, Berlin, 1961) *Theory of Diffraction* (Naval Intelligence Support Center, 1978)
18. V.I. Ivanov, Computation of corrections to the Fock asymptotics for the wave field near a circular cylinder and a sphere. J. Sov. Math. **20**(1), 1812–1817 (1982)
19. J.B. Keller, Diffraction by a convex cylinder. IRE Trans. Antennas Propag. **24**, 312–321 (1950)
20. J.B. Keller, Geometrical theory of diffraction. J. Opt. Soc. Am. **52**, 116–130 (1962)
21. N.Ya. Kirpichnikova, M.M. Popov, The Leontovich-Fock parabolic equation method in problems of short-wave diffraction by prolate bodies. J. Math. Sci. **194**(1), 30–43 (2012)
22. M.A. Leontovich, On one method for the solution of the problem on electromagnetic waves propagation along the surface of the Earth. Izvestiya AN SSSR Phys. **8**(1), 16 (1944)
23. M.A. Leontovich, V.A. Fock, Solution of the problem of propagation of electromagnetic waves along the earth's surface by the method of parabolic equation. J. Phys. USSR **10**(1), 13 (1946)
24. G.D. Malyuzhinets, Inversion formula for the Sommerfeld integral. Sov. Phys. Doklady **3**, 52–56 (1958)
25. G.D. Malyuzhinets, Excitation, reflection and radiation of surface waves from a wedge with a given surface impedances. Sov. Phys. Doklady **3**, 752–755 (1958)
26. F. Molinet, I.V. Andronov, D. Bouche, *Asymptotic and Hybrid Methods in Electromagnetics* (IEE, London, 2005)
27. B. Nobl, *Methods Based on the Wiener-Hopf Technique for the Solution of Partial Differential Equations* (Pergamon Press, 1958)
28. J.A. Stratton, L.J. Chu, Diffraction theory of electromagnetic waves. Phys. Rev. **56**(1), 99–107 (1939)

Chapter 2
Diffraction by an Elliptic Cylinder

2.1 Introduction

In this chapter, we consider problems of diffraction by an elliptic cylinder. It is worth noting that the problem of diffraction by an elliptical cylinder is of continuous interest. Starting with the works of Lord Rayleigh [23] up to modern research, the elliptical cylinder geometry is taken as the model for the study of different physical effects. This is because of the fact that, on the one hand, geometry is characterized by variable curvature of the surface which is controlled by a parameter that allows variation of the geometry from one limiting case (the strip) to another limiting case (the circular cylinder), and, on the other hand, this is a canonical problem that allows explicit solution by means of variable separation. The classical results of [11, 17, 18, 20, 21] and others are summarized in [12]. The analytical solutions are expressed in the form of series in terms of Mathieu functions. These series are suitable for calculations if the cylinder cross-section has a small wave size. For higher frequencies, one needs to use asymptotic techniques. The case, when the ellipse in the cross-section of the cylinder has a large aspect ratio, faces additional difficulties associated with the stability of the calculations. If the ellipse is sufficiently much elongated, there may appear a gap between the frequency range where numerical calculation can be performed and the frequency range for which the asymptotic formulas of classical diffraction theory provide acceptable accuracy. The asymptotic approximation discussed in this chapter is intended to bridge this gap. We present the results published in [2–4, 8, 9].

So, in this chapter we consider problems of diffraction by an elliptic cylinder with a strongly elongated cross-section. The incident fields are supposed not to depend on the coordinate directed along the axis of the cylinder. Therefore, the problems are essentially two-dimensional. Therefore, there the electromagnetic problems are reduced to scalar ones, and we do not distinguish the case of the acoustic field diffraction by a soft surface with the TE electromagnetic wave diffraction (in this case, u stands for the acoustic field or for the transverse electric component of the electromagnetic field). Similarly, the acoustic field diffraction by a hard surface is

I. Andronov, *Problems of High Frequency Diffraction by Elongated Bodies*, Springer Series in Optical Sciences 243, https://doi.org/10.1007/978-981-99-1276-6_2

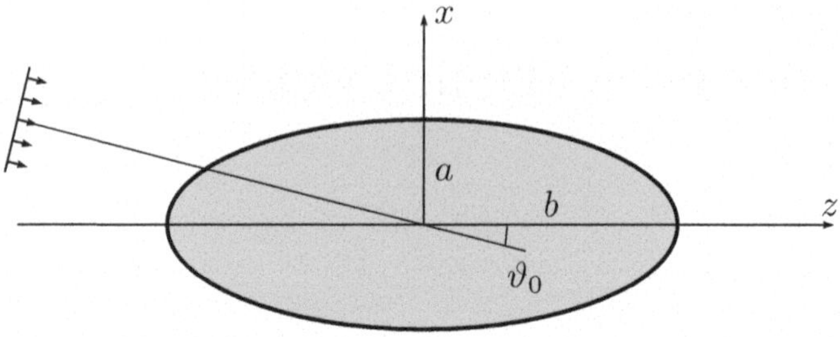

Fig. 2.1 Geometry of the problem

described by the same boundary-value problem as the TM electromagnetic wave diffraction, thus in that case u stands for the acoustic field or for the transverse magnetic component of the electromagnetic field. In both cases, u satisfies the Helmholtz equation

$$\Delta u + k^2 u = 0. \tag{2.1}$$

We denote the major semiaxis of the ellipse being the cross-section of the cylinder as b, while the minor semiaxis is denoted as a. The geometry of the problem is shown in Fig. 2.1. Thus, the surface of the cylinder is given by the equation

$$\frac{z^2}{b^2} + \frac{x^2}{a^2} = 1.$$

We consider high frequencies, such that $kb \gg 1$. However, it is more convenient to use kp, where $p = sqrtb^2 - a^2$ is the half focal distance, as the main asymptotic parameter. Besides, we assume that there is another large parameter in the problem. This is the aspect ratio of the semiaxes of the ellipse, i.e. b/a. So, there are two large parameters kp and b/a. It is important to fix the relation between these two parameters. In the frame of the asymptotic theory for strongly elongated bodies, we consider the case, when these two parameters are related to each other as

$$\chi \equiv kp\left(\frac{b}{a}\right)^{-2} \approx \frac{ka^2}{p}$$

and assume χ to be bounded while $kp \to \infty$. The parameter χ characterizes the rate of elongation of the cross-section of the cylinder. When χ decreases, the ellipse becomes more elongated. Parameter χ is well known for the problems of diffraction by an ellipse. In [14] basing on the exact solution of [17], the asymptotic series with respect to $\sqrt{\chi}$ was constructed for the case of $\chi \ll 1$.

We also restrict the possible directions of incidence by the requirement of the paraxial regime. That is the main direction of propagation of the incident wave which should be almost parallel to the major axis of the ellipse. This requirement is specified more precisely below. For the case of the plane wave incidence at the angle ϑ_0, it coincides with (1.16)

$$|\vartheta_0|\sqrt{kp} < const.$$

2.2 Stretched Coordinates and Separation of Variables

Elliptic coordinates (ξ, η) can be introduced by the usual formulas [12, 19]

$$z = p\eta\xi, \qquad x = p\sqrt{1-\eta^2}\sqrt{\xi^2 - 1}. \tag{2.2}$$

Coordinates vary so that $\xi \geq 1$ and $\eta \in [-1, 1]$. That means that only positive values of x coordinates are possible in (2.2). To overcome this, we split the field to the even part, which we denote as u_e, and the odd part u_o, that is

$$u(z, x) = u_e(z, |x|) + \text{sign}(x)u_o(z, |x|). \tag{2.3}$$

Helmholtz equation (2.1) in elliptic coordinates reduces to

$$\sqrt{\xi^2-1}\frac{\partial}{\partial\xi}\left(\sqrt{\xi^2-1}\frac{\partial u}{\partial\xi}\right) + \sqrt{1-\eta^2}\frac{\partial}{\partial\eta}\left(\sqrt{1-\eta^2}\frac{\partial u}{\partial\eta}\right) + (kp)^2\left(\xi^2-\eta^2\right)u = 0. \tag{2.4}$$

In view of $b \gg a$, radial coordinate ξ takes the values close to one on the surface of the cylinder, and therefore we introduce the boundary-layer coordinate τ by stretching ξ such that

$$\xi = 1 + \frac{\chi}{2kp}\tau.$$

Up to smaller order terms in kp, the surface is defined by the equation $\tau = 1$. Coordinates (η, τ) play the role of the boundary-layer coordinates. One can compare the scales with the usual scales for the boundary-layer coordinates in the Fock domain (see formula 1.5) and find them coincidentally.

Further, following the parabolic equation method, we extract the exponential factor

$$u = \mathrm{e}^{ikp\eta}U(\eta, \tau). \tag{2.5}$$

Substituting from (2.5) to (2.4), we get the equation for the attenuation function U. In this equation, we sort the terms according to their order in the asymptotic parameter kp. Assuming that the derivatives with respect to the boundary-layer coordinates (η, τ) do not change the order of the terms, we get

$$\left(L_0 + \frac{1}{kp}L_1\right) U = 0, \tag{2.6}$$

where

$$L_0 = 4\tau\frac{\partial^2}{\partial\tau^2} + 2\frac{\partial}{\partial\tau} + 2i\chi(1-\eta^2)\frac{\partial}{\partial\eta} - i\chi\eta + \chi^2\tau,$$

$$L_1 = \chi\tau^2\frac{\partial^2}{\partial\tau^2} + \chi\tau\frac{\partial}{\partial\tau} + \chi(1-\eta^2)\frac{\partial^2}{\partial\eta^2} - \chi\eta\frac{\partial}{\partial\eta} + \frac{\chi^3}{4}\tau^2.$$

It is possible to search for the solution U of (2.6) in the form of the asymptotic series

$$U = \sum_{j=0}^{\infty} \frac{U_j(\eta,\tau)}{(kp)^j}.$$

This leads to the recurrent sequence of problems

$$L_0 U_j = -L_1 U_{j-1}, \qquad j = 0, 1, \dots$$

where we set $U_{-1} \equiv 0$. However in this chapter, we restrict the derivations to finding U_0 only, and below we suppress the subscript $_0$ on U. The leading order approximation U satisfies the equation

$$4\tau\frac{\partial^2 U}{\partial\tau^2} + 2\frac{\partial U}{\partial\tau} + 2\mathrm{i}\chi\left(1-\eta^2\right)\frac{\partial U}{\partial\eta} - \mathrm{i}\chi\eta U + \chi^2\tau U = 0. \tag{2.7}$$

As is usual in diffraction theory, we call this Shrödinger type equation as the parabolic equation. Compared to the Leontovich parabolic equation, which we considered in Chap. 1, this parabolic equation takes into account the variation of curvature of the surface.

2.3 The Forward Wave

2.3.1 Asymptotic Representation

Equation (2.7) allows variable separation. We substitute $F(\tau)G(\eta)$ instead of U and find the differential equations for $F(\tau)$

$$4\tau F'' + 2F' + \chi^2\tau F = -\mu F$$

and for $G(\eta)$

$$2\mathrm{i}\chi\left(1-\eta^2\right)G' - \mathrm{i}\chi\eta G = \mu G.$$

The first can be reduced to Whittaker equation [1] (see also Appendix A.2)

$$V''(\zeta) + \left(-\frac{1}{4} + \frac{it}{\zeta} + \frac{3}{16\zeta^2}\right) V(\zeta) = 0, \tag{2.8}$$

where we set $\mu = 4\chi t$. Solving the second equation, we get

$$G(\eta) = \frac{1}{\left(1-\eta^2\right)^{1/4}} \left(\frac{1-\eta}{1+\eta}\right)^{it}.$$

Thus, elementary solutions of (2.7) can be written in the form

$$U = (\chi\tau)^{-1/4} \left(1-\eta^2\right)^{-1/4} \left(\frac{1-\eta}{1+\eta}\right)^{it} V(-\mathrm{i}\chi\tau, t), \tag{2.9}$$

where V satisfies the Whittaker equation (2.8).

Below, we use three solutions of the Whittaker equation, which are characterized by their behavior at zero and infinity. The first two solutions are the regular Whittaker functions [1] (see also Appendix A.2 and the approximation (A.22) there)

$$M_{\mathrm{i}t,-1/4}(\zeta) \sim \zeta^{1/4}, \qquad z \to 0 \tag{2.10}$$

and

$$M_{\mathrm{i}t,1/4}(\zeta) \sim \zeta^{3/4}, \qquad z \to 0.$$

The third solution is the irregular Whittaker function

$$W_{\mathrm{i}t,-1/4}(\zeta) \equiv W_{\mathrm{i}t,1/4}(\zeta) \sim \zeta^{it} \mathrm{e}^{-\zeta/2}, \qquad |\zeta| \to +\infty, \quad \arg(\zeta) < \pi.$$

Functions with the second index $-1/4$ are convenient for writing the representation of the even part U_e of the field, while Whittaker functions with the second index $1/4$ will be used in the representation of the odd part U_o.

Choosing $V(-\mathrm{i}\chi\tau, t) = W_{\mathrm{i}t,1/4}(-\mathrm{i}\chi\tau)$ in (2.9) results in a solution which satisfies the radiation conditions. Therefore, the reflected field can be searched for in the form of the integral

$$U^{\mathrm{refl}}_{e,o} = (\chi\tau)^{-1/4}(1-\eta^2)^{-1/4} \int \left(\frac{1-\eta}{1+\eta}\right)^{it} B_{e,o}(t) W_{\mathrm{i}t,1/4}(-\mathrm{i}\chi\tau)\mathrm{d}t. \tag{2.11}$$

In view of $W_{\mathrm{i}t,-1/4}(z) = W_{\mathrm{i}t,1/4}(z)$, the representations for the even and odd parts coincide, however the amplitudes $B_e(t)$ and $B_o(t)$ are different. The path of integration will be chosen later; presently, we require only the convergence of the integrals and the possibility to differentiate the right-hand side of (2.11) under the sign of integration twice by τ and once by η.

The amplitudes $B_{e,o}(t)$ in (2.11) can be found then from the requirement that the total odd and even parts of the field satisfy the boundary conditions. For the case of the Dirichlet boundary condition, we get the integral equations

$$\int \left(\frac{1-\eta}{1+\eta}\right)^{it} B_{o,e}(t) W_{it,1/4}(-i\chi)dt = -\sqrt[4]{\chi}\sqrt[4]{1-\eta^2}e^{-ikp\eta}u^{\text{inc}}_{o,e}\Big|_{\tau=1}. \quad (2.12)$$

In the case of the Neumann boundary condition in view of

$$\frac{\partial}{\partial n}\Big|_{n=0} = \frac{2}{a\sqrt{1-\eta^2}}\frac{\partial}{\partial \tau}\Big|_{\tau=1},$$

we get

$$\int \left(\frac{1-\eta}{1+\eta}\right)^{it} B_{o,e}(t)\left(\frac{1}{4}W_{it,1/4}(-i\chi) + i\chi W'_{it,1/4}(-i\chi)\right) dt =$$
$$= \sqrt[4]{\chi}\sqrt[4]{1-\eta^2}e^{-ikp\eta}\frac{\partial u^{\text{inc}}_{o,e}}{\partial \tau}\Big|_{\tau=1}. \quad (2.13)$$

If we choose the path of integration in the representation (2.11) along the real axis of t, the integrals in (2.12) and (2.13) contain the integral transform (A.32). However, the integrals for the amplitudes $B_{o,e}$ given by the inversion of this integral transform appear too cumbersome, and it is difficult to reduce these integrals to special functions. Therefore, we use a different approach. Using (2.9), we represent the total field in the form of the integral with respect to the parameter of variable separation in (2.7), that is

$$U_{o,e} = (\chi\tau)^{-1/4}(1-\eta^2)^{-1/4}\int_{-\infty}^{+\infty}\left(\frac{1-\eta}{1+\eta}\right)^{it} A_{o,e}(t)\Big\{M_{it,s_{o,e}/4}(-i\chi\tau)$$
$$+ R_{o,e}(t)W_{it,1/4}(-i\chi\tau)\Big\}dt. \quad (2.14)$$

Here $s_e = -1$ and $s_o = 1$.

In the representation (2.14) compared to representation (2.11), we set $B_{o,e} = A_{o,e}R_{o,e}$. The terms containing Whittaker function M represent the incident field, and amplitudes A_e and A_o can be determined by equating expression (2.14) to the incident wave. The functions R_e and R_o can be chosen such that the boundary conditions are satisfied. It is sufficient to require that the boundary conditions are satisfied below the sign of integration, i.e. for any value of t. This yields for TE polarization (or acoustic waves on a soft surface)

$$R_{o,e}(t) = -\frac{M_{it,s_{o,e}/4}(-i\chi)}{W_{it,1/4}(-i\chi)}, \quad (2.15)$$

and for TM polarization (or acoustic waves on a hard surface)

$$R_{o,e}(t) = -\frac{4\mathrm{i}\chi \dot{M}_{\mathrm{i}t,s_{o,e}/4}(-\mathrm{i}\chi) + M_{\mathrm{i}t,s_{o,e}/4}(-\mathrm{i}\chi)}{4\mathrm{i}\chi \dot{W}_{\mathrm{i}t,1/4}(-\mathrm{i}\chi) + W_{\mathrm{i}t,1/4}(-\mathrm{i}\chi)}. \tag{2.16}$$

Thus, the problem of solving integral equations (2.12) and (2.13) is reduced to finding the amplitudes A_e and A_o.

2.3.2 *Integral Representation of the Incident Plane Wave*

For finding the amplitudes A_e and A_o in the representation (2.14), we match this expression with the incident field u^{inc}. First, we consider the case of plane wave incidence (see Fig. 2.1)

$$u^{\mathrm{inc}} = \mathrm{e}^{ikz\cos(\vartheta_0)+ikx\sin(\vartheta_0)}.$$

The plane wave u^{inc} is represented as the sum of its even and odd parts as in (2.3), where

$$u_e^{\mathrm{inc}} = \mathrm{e}^{ikz\cos(\vartheta_0)}\cos\left(kx\sin(\vartheta_0)\right), \quad u_o^{\mathrm{inc}} = \mathrm{i}\,\mathrm{e}^{ikz\cos(\vartheta_0)}\sin\left(kx\sin(\vartheta_0)\right). \tag{2.17}$$

Consider first the even part u_e^{inc}. In the boundary-layer coordinates (η, τ), it reduces, up to smaller order terms, to

$$u_e^{\mathrm{inc}} = \mathrm{e}^{ikp\eta}\exp\left(\frac{\mathrm{i}}{2}(\chi\tau - \alpha^2)\eta\right)\cos\left(\sqrt{1-\eta^2}\sqrt{\chi\tau}\alpha\right).$$

Here, $\alpha = \vartheta_0(kp)^{-1/2}$ is assumed to be of the order of unity in agreement with the paraxial incidence assumption (1.16).

Equating expression (2.17) to the representation (2.14) yields the integral equation for the amplitude $A_e(t)$

$$(\chi\tau)^{-1/4}(1-\eta^2)^{-1/4}\int_{-\infty}^{+\infty}\left(\frac{1-\eta}{1+\eta}\right)^{\mathrm{i}t} A_e(t) M_{\mathrm{i}t,-1/4}(-\mathrm{i}\chi\tau)\mathrm{d}t =$$

$$= \exp\left(\frac{\mathrm{i}}{2}\left(\chi\tau - \alpha^2\right)\eta\right)\cos\left(\sqrt{1-\eta^2}\sqrt{\chi\tau}\alpha\right). \tag{2.18}$$

This integral equation is not simpler than (2.12), but it contains parameters, the arbitrariness of which can be exploited.

We note that the elongation parameter χ is presented in (2.18) only in the form of the product $\chi\tau$. Therefore, as A_e does not depend on τ, it either does not depend on χ. Further, $\chi\tau$ and α^2 are presented on the right-hand side of (2.18) in a symmetric manner. Letting

$$A_e(t) = \frac{M_{\mathrm{i}t,-1/4}(\mathrm{i}\alpha^2)}{\sqrt{|\alpha|}} a_e(t), \tag{2.19}$$

we conclude that $a_e(t)$ does not depend on α either.

We also note that both the left-hand side of (2.18) and its right-hand side satisfy the parabolic equation (2.7), and, therefore, it is sufficient to solve (2.19) for any fixed values of τ and α. Letting $\chi\tau$ and α infinitely close to zero and using the asymptotics (2.10) simplifies the integral equation for $a_e(t)$ to

$$\int\limits_{-\infty}^{+\infty} a_e(t) \left(\frac{1-\eta}{1+\eta}\right)^{it} \mathrm{d}t = (1-\eta^2)^{1/4}.$$

Now we use the inverse transform (A.33), which yields

$$a_e(t) = \frac{1}{\pi} \int\limits_0^{+\infty} \left(\frac{1+\eta}{1-\eta}\right)^{it} \left(1-\eta^2\right)^{-3/4} \mathrm{d}\eta. \tag{2.20}$$

The integral in (2.20) can be expressed in terms of gamma-functions. For that, we change the integration variable to $s = (1+\eta)/2$ and get

$$a_e(t) = \frac{1}{\pi\sqrt{2}} \int\limits_0^1 s^{\mathrm{i}t-3/4}(1-s)^{-\mathrm{i}t-3/4} \mathrm{d}s.$$

The integral now coincides with the integral representation for the Beta function, which yields

$$a_e(t) = \frac{1}{\sqrt{2}\pi^{3/2}} \Gamma\left(\frac{1}{4} + \mathrm{i}t\right) \Gamma\left(\frac{1}{4} - \mathrm{i}t\right).$$

Finally from (2.19), we get

$$A_e(t) = \frac{1}{\sqrt{2}\pi^{3/2}} \frac{M_{\mathrm{i}t,-1/4}(\mathrm{i}\alpha^2)}{\sqrt{|\alpha|}} \Gamma\left(\frac{1}{4} + \mathrm{i}t\right) \Gamma\left(\frac{1}{4} - \mathrm{i}t\right). \tag{2.21}$$

Consider now the odd part u_o^{inc}. Retaining only the leading order terms, we get the integral equation

$$(\chi\tau)^{-1/4}(1-\eta^2)^{-1/4}\int\limits_{-\infty}^{+\infty}\left(\frac{1-\eta}{1+\eta}\right)^{it} A_o(t)M_{\mathrm{i}t,1/4}(-i\chi\tau)dt =$$

$$= \mathrm{i}\,\exp\left(\frac{\mathrm{i}}{2}\left(\chi\tau-\alpha^2\right)\eta\right)\sin\left(\sqrt{1-\eta^2}\sqrt{\chi\tau}\alpha\right).$$

Performing similar derivations as in the case of (2.18), we find

$$A_o(t) = \mathrm{sign}(\alpha)\frac{2^{3/2}\mathrm{i}}{\pi^{3/2}}\frac{M_{\mathrm{i}t,1/4}(\mathrm{i}\alpha^2)}{\sqrt{|\alpha|}}\Gamma\left(\frac{3}{4}+\mathrm{i}t\right)\Gamma\left(\frac{3}{4}-\mathrm{i}t\right). \tag{2.22}$$

2.3.3 *Integral Representation of the Line Source Field*

Another possible incident field is the field of the line source

$$u^{\mathrm{inc}} = -\frac{i}{4}H_0^{(1)}(kr), \qquad r = \sqrt{(z-z_0)^2+(x-x_0)^2}, \tag{2.23}$$

where $H_0^{(1)}$ is the Hankel function of the first kind [1]. This incident field is the 2D Green's function corresponding to the point source at point (z_0, x_0). We assume that the source is located on the left of the ellipse, i.e. $z_0 < -b$, and that everywhere in the boundary layer near the surface the quantity kr is large. Besides, the direction of incidence should be at a small angle to the axis of the ellipse. That results in the following restrictions for the coordinates of the source

$$z_0 < -b, \qquad k(-b-z_0) \gg 1, \qquad \left|\frac{x_0}{z_0}\right|\sqrt{kp} \le const.$$

The structure of the field representation (2.14) is such that only the amplitudes $A_e(t)$ and $A_o(t)$ presented in it depend on the incident field. Therefore, we only need to find these amplitude functions in accordance with the incident field given by (2.23).

Thus, we search for the representation of the incident field (2.23) in the form of the integrals

$$U_{o,e}^{\mathrm{inc}} = (\chi\tau)^{-1/4}(1-\eta^2)^{-1/4}\int\limits_{-\infty}^{+\infty}\left(\frac{1-\eta}{1+\eta}\right)^{\mathrm{i}t} A_{o,e}(t)M_{\mathrm{i}t,s_{o,e}/4}(-\mathrm{i}\chi\tau)\mathrm{d}t. \tag{2.24}$$

We remind that $s_e = -1$ and $s_o = 1$. For that, we need first to find the expression for the incident field u^{inc} in terms of the boundary-layer coordinates (η, τ) and neglect asymptotically small terms. To find this expression, we replace the Hankel function in (2.23) by its asymptotic representation [1]

$$u^{\text{inc}} \sim -\frac{\mathrm{i}}{4}\sqrt{\frac{2}{\pi kr}}\mathrm{e}^{\mathrm{i}kr-\mathrm{i}\pi/4}.$$

Further, we expand the distance r as

$$r = (z - z_0) + \frac{1}{2}\frac{(x - x_0)^2}{z - z_0} + \dots$$

and substitute here the expressions for z and x via the boundary-layer coordinates

$$z = p\eta + \frac{\chi}{2k}\eta\tau, \qquad x = p\sqrt{1-\eta^2}\sqrt{\frac{\chi}{kp}}\tau + \dots$$

Neglecting asymptotically small terms in the expression for the distance r, we get the approximations for the attenuation functions

$$U_e^{\text{inc}} = \frac{C}{\sqrt{\eta+\zeta_0}}\exp\left(\frac{\mathrm{i}\chi}{2}\frac{(1+\eta\zeta_0)\tau + \gamma_0^2}{\eta+\zeta_0}\right)\cos\left(\frac{\chi\gamma_0\sqrt{1-\eta^2}\sqrt{\tau}}{\eta+\zeta_0}\right), \tag{2.25}$$

$$U_o^{\text{inc}} = \frac{\mathrm{i}C}{\sqrt{\eta+\zeta_0}}\exp\left(\frac{\mathrm{i}\chi}{2}\frac{(1+\eta\zeta_0)\tau + \gamma_0^2}{\eta+\zeta_0}\right)\sin\left(\frac{\chi\gamma_0\sqrt{1-\eta^2}\sqrt{\tau}}{\eta+\zeta_0}\right). \tag{2.26}$$

Here, we introduced scaled coordinates of the source

$$\zeta_0 = -\frac{z_0}{b}, \qquad \gamma_0 = \frac{x_0}{a}$$

and the coefficient

$$C = -\frac{\mathrm{e}^{\mathrm{i}\pi/4}}{4}\sqrt{\frac{2}{\pi}}\frac{\mathrm{e}^{-\mathrm{i}kz_0}}{\sqrt{kb}}.$$

It is easy to check that the right-hand sides of (2.25) and (2.26) satisfy parabolic equation (2.7). So, we equate representation (2.24) for the even part of the incident field with (2.25) and for the odd part of the field with (2.26). This yields integral equations for the amplitudes A_e and A_o. These equations contain τ as a parameter and the solutions do not depend on τ. This enables us to simplify the equations by setting τ infinitely small. With the use of the asymptotics (2.10), we get

$$\int_{-\infty}^{+\infty}\left(\frac{1-\eta}{1+\eta}\right)^{\mathrm{i}t} A_e(t)\mathrm{d}t = C\mathrm{e}^{\mathrm{i}\pi/8}\frac{\sqrt[4]{1-\eta^2}}{\sqrt{\eta+\zeta_0}}\exp\left(\frac{\mathrm{i}\chi}{2}\frac{\gamma_0^2}{\eta+\zeta_0}\right), \tag{2.27}$$

$$\int_{-\infty}^{+\infty} \left(\frac{1-\eta}{1+\eta}\right)^{it} A_o(t)\mathrm{d}t = -C\mathrm{e}^{3\mathrm{i}\pi/8}\sqrt{\chi}\gamma_0 \frac{(1-\eta^2)^{3/4}}{(\eta+\zeta_0)^{3/2}} \exp\left(\frac{\mathrm{i}\chi}{2}\frac{\gamma_0^2}{\eta+\zeta_0}\right). \quad (2.28)$$

Integral equations (2.27) and (2.28) can be solved with the help of the transform (A.32) and (A.33). The trick with letting $\tau \to 0$ makes the integrals much simpler and enables us to express amplitudes $A_e(t)$ and $A_o(t)$ in terms of Whittaker functions.

Consider the first integral equation (2.27). Its solution can be written as

$$A_e(t) = \frac{C\mathrm{e}^{\mathrm{i}\pi/8}}{\pi} \int_{-1}^{1} \left(\frac{1+\eta}{1-\eta}\right)^{it} \exp\left(\frac{\mathrm{i}\chi}{2}\frac{\gamma_0^2}{\eta+\zeta_0}\right) \frac{\mathrm{d}\eta}{\sqrt{\eta+\zeta_0}(1-\eta^2)^{3/4}}.$$

First, we change the integration variable to $s_1 = (1+\eta)/(1-\eta)$ and get

$$A_e(t) = \frac{C\mathrm{e}^{\mathrm{i}\pi/8}}{\pi\sqrt{2}} \frac{\exp\left(-\frac{\mathrm{i}\chi\rho_0^2}{2}\frac{1}{\zeta_0-1}\right)}{\sqrt{\zeta_0-1}} \int_0^{+\infty} \frac{s_1^{it-3/4}}{\sqrt{\theta s_1+1}} \exp\left(\frac{\mathrm{i}\chi\gamma_0^2}{\zeta_0^2-1}\frac{1}{\theta s_1+1}\right) \mathrm{d}s_1,$$

where $\theta = (\zeta_0+1)/(\zeta_0-1)$. Further, introducing the integration variable $s_2 = (1-\theta s_1)/(1+\theta s_1)$ yields

$$A_e - \frac{C\mathrm{e}^{\mathrm{i}\pi/8}}{\pi} \frac{1}{\sqrt[4]{\zeta_0^2-1}} \left(\frac{\zeta_0-1}{\zeta_0+1}\right)^{it} \exp\left(\frac{\mathrm{i}\chi}{2}\frac{\gamma_0^2\zeta_0}{\zeta_0-1}\right) \times$$

$$\times \int_{-1}^{1} (1-s_2)^{\mathrm{i}t-3/4}(1+s_2)^{-\mathrm{i}t-3/4} \exp\left(\frac{\mathrm{i}}{2}\Upsilon_0 s_2\right) \mathrm{d}s_2, \quad (2.29)$$

where

$$\Upsilon_0 = \frac{\chi\gamma_0^2}{\zeta_0^2-1}.$$

Comparing the integral in (2.29) with the integral representation (A.9) for the Whittaker function M, we finally get

$$A_e = -\frac{\mathrm{e}^{\mathrm{i}\pi/4}}{4\pi^2} \frac{\mathrm{e}^{-ikz_0}}{\sqrt{kb}} \frac{\mathrm{e}^{\mathrm{i}\Upsilon_0\zeta_0/2}}{\sqrt[4]{\zeta_0^2-1}} \left(\frac{\zeta_0-1}{\zeta_0+1}\right)^{it} \times$$

$$\times \Gamma\left(\frac{1}{4}+\mathrm{i}t\right) \Gamma\left(\frac{1}{4}-\mathrm{i}t\right) \frac{M_{\mathrm{i}t,-1/4}(\mathrm{i}\Upsilon_0)}{\sqrt[4]{\Upsilon_0}}. \quad (2.30)$$

We perform analogous transformations with the integral representation for the solution of the integral equation (2.28), which result in the expression

$$A_o = \operatorname{sign}(x_0)\frac{\mathrm{e}^{-\mathrm{i}\pi/4}}{\pi^2}\frac{\mathrm{e}^{-ikz_0}}{\sqrt{kb}}\frac{\mathrm{e}^{\mathrm{i}\Upsilon_0\zeta_0/2}}{\sqrt[4]{\zeta_0^2-1}}\left(\frac{\zeta_0-1}{\zeta_0+1}\right)^{it}\times$$

$$\times\,\Gamma\left(\frac{3}{4}+\mathrm{i}t\right)\Gamma\left(\frac{3}{4}-\mathrm{i}t\right)\frac{M_{\mathrm{i}t,1/4}(\mathrm{i}\Upsilon_0)}{\sqrt[4]{\Upsilon_0}}. \qquad (2.31)$$

Comparing expressions (2.30) and (2.31) with (2.21) and (2.22), we can find out that the amplitudes $A_{o,e}$ in the case of the line source differ from that in the case of plane wave incidence by the replacement of α^2 with Υ_0 and by the additional factor

$$\Omega\left(\frac{\zeta_0-1}{\zeta_0+1}\right)^{\mathrm{i}t},$$

where

$$\Omega = -\frac{\mathrm{e}^{\mathrm{i}\pi/4}}{2^{3/2}\sqrt{\pi}}\frac{\mathrm{e}^{-\mathrm{i}kz_0+\mathrm{i}\Upsilon_0\zeta_0/2}}{\sqrt{kb}\sqrt[4]{\zeta_0^2-1}} \qquad (2.32)$$

does not depend on the integration variable.

2.4 The Backward Wave

In the previous section, we have derived the leading order approximation for the field of diffraction by a strongly elongated ellipse. This approximation represents the wave which propagates in the positive direction of η along the surface with the main factor $\mathrm{e}^{\mathrm{i}kp\eta}$. However, the actual electromagnetic or acoustic field consists of two wave processes. One runs in the forward direction and the other, originated, when the forward wave encircles the shadowed extremity of the ellipse, runs in the backward direction. In order to get an accurate approximation of the field in the boundary layer near the surface, this backward wave contribution should be taken into account.

2.4.1 *Integral Representation for the Backward Wave*

In the parabolic equation method for the backward wave, we extract the factor $\mathrm{e}^{-\mathrm{i}kp\eta}$ and come to the parabolic equation which differs from (2.7) only in the sign at the derivative by η. Therefore, the representation of the backward wave field can be obtained by simply inverting η coordinate in (2.14). Taking into account that there is no backward incident wave, we immediately get

$$u_{o,e}^{\text{back}} = \frac{\mathrm{e}^{-\mathrm{i}kp\eta}}{(\chi\tau)^{1/4}(1-\eta^2)^{1/4}} \int\limits_{-\infty}^{+\infty} \left(\frac{1-\eta}{1+\eta}\right)^{\mathrm{i}t} B_{o,e}(t) r_{o,e}(t) W_{\mathrm{i}t,1/4}(-\mathrm{i}\chi\tau)\mathrm{d}t. \quad (2.33)$$

Here, we represented the amplitudes of the backward wave process as the products of the forward wave process amplitude $B_{o,e}$ and an additional reflection coefficient $r_{o,e}(t)$ which can be thought of as the reflection coefficient from the shadowed extremity of the elliptic cylinder. To find $r_{o,e}$, one needs to consider the fields in the vicinity of the shadowed extremity of the ellipse and match the representation of the field there with the sum of the forward and backward wave processes.

2.4.2 *The Reflection Coefficients*

The difficulty in the case of strongly elongated geometries is in the absence of the large asymptotic parameter in the domain near the shadowed extremity of the body. Indeed, the radius of curvature at the trailing end is $\rho_e = a^2/b$, which means that $k\rho_e \approx \chi$ and therefore is a quantity of order one. Nevertheless, there is a way out and it is in the possibility to approximate an elongated ellipse with a parabola near its tip. Therefore, we introduce parabolic coordinates (v, w)

$$z = p + \frac{v-w}{2k}, \qquad x = \frac{\sqrt{vw}}{k},$$

in terms of which the surface is specified by $v = \chi$.

The parabolic coordinates allow the separation of variables directly in Helmholtz equation (2.1) which reduces to

$$4v\frac{\partial^2 U}{\partial v^2} + 2\frac{\partial U}{\partial v} + vU + 4w\frac{\partial^2 U}{\partial w^2} + 2\frac{\partial U}{\partial w} + wU = 0.$$

Therefore, the solution can be represented as the integral

$$u_{o,e} = \frac{1}{\sqrt[4]{vw}} \int P_{o,e}(t) W_{\mathrm{i}t,1/4}(-\mathrm{i}v) M_{\mathrm{i}t,s_{o,e}/4}(\mathrm{i}w)\mathrm{d}t. \quad (2.34)$$

We have chosen the Whittaker function depending on w in these solutions in agreement with the symmetry of the field with respect to the x coordinate. The choice of the Whittaker function depending on v is governed by the radiation condition. The amplitudes $P_{0,e}$ should be found, when (2.34) is matched with the incoming forward wave (2.14).

The representation (2.14), as well as (2.33) for the backward wave, is valid only in the middle part of the elliptic cylinder. When approaching the ends of the ellipse, that is when $\eta \to \pm 1$, it may diverge. The representation (2.33) is on the contrary valid near the end of the ellipse when w is not large. There is an intermediate domain, such that η is not too close to 1 and w is not too large. In this domain, we perform the matching of the two asymptotic expressions. For that in solution (2.34), we replace Whittaker function $M_{\mathrm{i}t,s_{o,e}/4}(\mathrm{i}w)$ with its asymptotics (A.24):

$$M_{\mathrm{i}t,\ell}(\mathrm{i}w) \sim \exp\left(-\frac{\mathrm{i}w}{2} + \mathrm{i}\pi\left(\ell + \frac{1}{2} - \mathrm{i}t\right)\right) \frac{\Gamma(2\ell+1)}{\Gamma\left(\ell + \frac{1}{2} + \mathrm{i}t\right)} w^{\mathrm{i}t} \mathrm{e}^{-\pi t/2} + \exp\left(\frac{\mathrm{i}w}{2}\right) \frac{\Gamma(2\ell+1)}{\Gamma\left(\ell + \frac{1}{2} - \mathrm{i}t\right)} w^{-\mathrm{i}t} \mathrm{e}^{\pi t/2}. \quad (2.35)$$

Noting the relation between the parabolic coordinates (v, w) and the elliptic coordinates (η, τ), expressed at the leading order by formulas

$$\eta = 1 - \frac{w}{2kp}, \qquad \tau = \frac{v}{\chi},$$

we conclude that the dependencies on v and τ coincide at the leading order, and it is sufficient to equate the integrands only on the surface. The first term of (2.35), when substituted to (2.34), matches the incoming (forward) wave, and after simple derivations we get

$$P_{o,e} = \sqrt[4]{kp}\, B_{o,e}(t) \exp\left(\mathrm{i}kp - s_{o,e}\frac{\mathrm{i}\pi}{4} - \frac{\pi t}{2}\right) \left(4kp\right)^{-\mathrm{i}t} \frac{\Gamma\left(\frac{2-s_{o,e}}{4} + \mathrm{i}t\right)}{\Gamma\left(\frac{2-s_{o,e}}{2}\right)}. \quad (2.36)$$

The second term of (2.35) matches the backward wave, and we find

$$r_{o,e}(t) = \exp\left(2\mathrm{i}kp - s_{o,e}\frac{\mathrm{i}\pi}{4}\right) \left(4kp\right)^{-2\mathrm{i}t} \frac{\Gamma\left(\frac{2-s_{o,e}}{4} + \mathrm{i}t\right)}{\Gamma\left(\frac{2-s_{o,e}}{4} - \mathrm{i}t\right)}. \quad (2.37)$$

It is worth noting that the reflection coefficients $r_{o,e}$ contain asymptotic parameter kp, firstly in the exponential, which describes the phase shift corresponding to the path from the light-shadow boundary to the shaded tip and back, and secondly in the multiplier

$$(kp)^{-2\mathrm{i}t} = \exp\Big(-2\mathrm{i}\log(kp)t\Big).$$

This multiplier causes the attenuation of the field. Compared to the usual creeping waves, this attenuation is only logarithmically large. Generally speaking, the representation for the backward field allows further asymptotic simplifications, being in the computation of the integral by means of the residue theorem. However, we do not perform this reduction because finding the poles in the lower half-plane of the integration variable t is not easy (see, however, Sect. A.3 where these poles are studied).

Besides, there is no good representation for the derivative of Whittaker functions with respect to the first index, which makes it possible to use only numerical differentiation resulting in the loss of accuracy. Finally, as we will see below, the asymptotic expression for the total field expressed as the sum of (2.14) and (2.33) provides a sufficiently accurate approximation even at $kb \sim 2$, for which the approximation with one or two residues evidently fails.

2.5 The Induced Currents

2.5.1 Integral Representations for the Currents

For the TE case, the currents run in transverse direction $\mathbf{e}_y$ and are given by

$$J_{TE} = \frac{1}{k}\frac{\partial u}{\partial n}\bigg|_{n=1} = \frac{2}{ka\sqrt{1-\eta^2}}\frac{\partial u}{\partial \tau}\bigg|_{\tau=1}.$$

When computing the derivative of the right-hand side of (2.14) with respect to τ, we note that the differentiation of the multiplier outside the integral leads to the term, which vanishes at $\tau = 1$. So, it is necessary to differentiate only the terms in curly braces. This results in the expression

$$\frac{\partial}{\partial \tau}\{\ldots\}\bigg|_{\tau=1} = \mathrm{i}\chi\left(\frac{M_{\mathrm{i}t,s_{o,e}/4}(-\mathrm{i}\chi)}{W_{\mathrm{i}t,1/4}(-\mathrm{i}\chi)}\dot{W}_{\mathrm{i}t,1/4}(-I\chi) - \dot{M}_{\mathrm{i}t,s_{o,e}/4}(-\mathrm{i}\chi)\right)$$

which after bringing the terms to a common denominator reduces to

$$\frac{\partial}{\partial \tau}\{\ldots\}\bigg|_{\tau=1} = -\mathrm{i}\chi\frac{\mathcal{W}_{\mathrm{i}t,s_{o,e}/4}}{W_{\mathrm{i}t,1/4}(-\mathrm{i}\chi)},$$

where the Wronskian $\mathcal{W}$ of Whittaker functions M and W is given by (A.21)

$$\mathcal{W}_{\mathrm{i}t,\ell} = \dot{M}_{\mathrm{i}t,\ell}W_{\mathrm{i}t,\ell} - M_{\mathrm{i}t,\ell}\dot{W}_{\mathrm{i}t,\ell} = \frac{\Gamma(2\ell)}{\Gamma(\ell+1/2-\mathrm{i}t)}. \tag{2.38}$$

For the current of the backward wave, we rewrite the expression

$$\frac{M_{\mathrm{i}t,s_{o,e}/4}(-\mathrm{i}\chi)}{W_{\mathrm{i}t,1/4}(-\mathrm{i}\chi)}\dot{W}_{\mathrm{i}t,1/4}(-\mathrm{i}\chi),$$

which appears after differentiating (2.33), as

$$\dot{M}_{\mathrm{i}t,s_{o,e}/4}(-\mathrm{i}\chi) - \frac{\mathcal{W}_{\mathrm{i}t,s_{o,e}/4}}{W_{\mathrm{i}t,1/4}(-\mathrm{i}\chi)}$$

and exclude the first term, which has no singularities in the upper half-plane of t and therefore gives no contribution to the integral.

Finally, combining the currents of the forward and the backward waves, we get

$$J^{TE}_{o,e} = \frac{8\mathrm{i}\chi^{3/4}\Gamma(s_{o,e}/2)}{ka\left(1-\eta^2\right)^{3/4}} \int_{-\infty}^{+\infty} \frac{A_{o,e}(t)}{W_{\mathrm{i}t,1/4}(-\mathrm{i}\chi)\Gamma(1/2+s_{o,e}/4-\mathrm{i}t)} \times$$
$$\times \left\{ \left(\frac{1-\eta}{1+\eta}\right)^{\mathrm{i}t} \mathrm{e}^{\mathrm{i}kp\eta} + \left(\frac{1+\eta}{1-\eta}\right)^{\mathrm{i}t} \mathrm{e}^{-\mathrm{i}kp\eta} r_{o,e}(t) \right\} \mathrm{d}t. \quad (2.39)$$

For the case of TM polarization, the currents are directed along $\mathbf{e}_\eta$ and coincide with H_y component of the magnetic field on the surface. That is $J^{TM} = u|_{\tau=1}$. After setting $\tau = 1$ in (2.14) and bringing the terms in curly braces to the common denominator, we get

$$U_{o,e}|_{\tau=1} = -\frac{4\mathrm{i}\chi^{3/4}}{\sqrt[4]{1-\eta^2}} \int_{-\infty}^{+\infty} \left(\frac{1-\eta}{1+\eta}\right)^{\mathrm{i}t} \frac{A_{o,e}(t)\, \mathcal{W}_{\mathrm{i}t,s_{o,e}/4}\, \mathrm{d}t}{4\mathrm{i}\chi \dot{W}_{\mathrm{i}t,1/4}(-\mathrm{i}\chi) + W_{\mathrm{i}t,1/4}(-\mathrm{i}\chi)},$$

where the Wronskian $\mathcal{W}_{\mathrm{i}t,s_{o,e}/4}$ is given by (2.38).

In the approximation for the backward wave, we use the relation

$$\dot{M}_{\mathrm{i}t,s_{o,e}}(-\mathrm{i}\chi) W_{\mathrm{i}t,1/4}(-\mathrm{i}\chi) = \mathcal{W}_{\mathrm{i}t,s_{o,e}/4} + M_{\mathrm{i}t,s_{o,e}/4}(-\mathrm{i}\chi) \dot{W}_{\mathrm{i}t,1/4}(-\mathrm{i}\chi),$$

which yields

$$R_{o,e} W_{\mathrm{i}t,s_{o,e}/4}(-\mathrm{i}\chi) = \frac{\mathcal{W}[M_{\mathrm{i}t,s_{o,e}/4}, W_{\mathrm{i}t,s_{o,e}/4}]}{4\mathrm{i}\chi \dot{W}_{\mathrm{i}t,1/4}(-\mathrm{i}\chi) + W_{\mathrm{i}t,1/4}(-\mathrm{i}\chi)} - M_{\mathrm{i}t,s_{o,e}/4}(-\mathrm{i}\chi).$$

The last term can be thrown away because it has no singularities in the upper half-plane of t. Thus, we get

$$J^{TM}_{o,e} = -\frac{4\mathrm{i}\chi^{3/4}\Gamma(s_{o,e}/2)}{\sqrt[4]{1-\eta^2}} \int_{-\infty}^{+\infty} \frac{A_{o,e}(t)}{4\mathrm{i}\chi \dot{W}_{\mathrm{i}t,1/4}(-\mathrm{i}\chi) + W_{\mathrm{i}t,1/4}(-\mathrm{i}\chi)} \times$$
$$\times \left\{ \left(\frac{1-\eta}{1+\eta}\right)^{\mathrm{i}t} \mathrm{e}^{\mathrm{i}kp\eta} + \left(\frac{1+\eta}{1-\eta}\right)^{\mathrm{i}t} \mathrm{e}^{-\mathrm{i}kp\eta} r_{o,e}(t) \right\} \frac{\mathrm{d}t}{\Gamma(1/2+s_{o,e}/4-\mathrm{i}t)}. \quad (2.40)$$

In the expressions (2.39) and (2.40), one needs to substitute the expressions for the amplitudes A_o and A_e, which depend on the incident wave.

2.5.2 The Case of Plane Wave Incidence

It may be convenient to express Whittaker functions via Coulomb wave functions F and H^+ [1] in the above expressions. Using the formulas from Appendix A.2.2 after cumbersome, but straightforward, derivations, we get the leading order approximations for the induced currents in the case of plane wave incidence

$$J_{TE} \sim \frac{1}{\sqrt{kp}} \frac{2\sqrt{2}\,\chi^{1/4}}{(1-\eta)^{3/4}\sqrt{\pi\alpha}} \int\limits_{-\infty}^{+\infty} \sum_{j=0,1} \frac{(\mathrm{i}\,\mathrm{sign}(\vartheta_0 x))^j\; F_{\frac{2j-3}{4}}\left(t, \frac{\alpha^2}{2}\right)}{H^+_{\frac{2j-3}{4}}\left(-t, \frac{\chi}{2}\right)} \times \\ \times \left\{ \left(\frac{1-\eta}{1+\eta}\right)^{\mathrm{i}t} \mathrm{e}^{\mathrm{i}kp\eta} + r_j(t) \left(\frac{1+\eta}{1-\eta}\right)^{\mathrm{i}t} \mathrm{e}^{-\mathrm{i}kp\eta} \right\} \mathrm{d}t, \quad (2.41)$$

$$J_{TM} \sim \frac{4\sqrt{2}\,\chi^{3/4}}{\sqrt[4]{1-\eta^2}\sqrt{\pi\alpha}} \int\limits_{-\infty}^{+\infty} \sum_{j=0,1} \frac{(\mathrm{i}\,\mathrm{sign}(\vartheta_0 x))^j\; F_{-\frac{3-2j}{4}}\left(t, \frac{\alpha^2}{2}\right)}{H^+_{\frac{2j-3}{4}}\left(-t, \frac{\chi}{2}\right) - 2\chi \dot{H}^+_{\frac{2j-3}{4}}\left(-t, \frac{\chi}{2}\right)} \times \\ \times \left\{ \left(\frac{1-\eta}{1+\eta}\right)^{\mathrm{i}t} \mathrm{e}^{\mathrm{i}kp\eta} + r_j(t) \left(\frac{1+\eta}{1-\eta}\right)^{\mathrm{i}t} \mathrm{e}^{-\mathrm{i}kp\eta} \right\} \mathrm{d}t. \quad (2.42)$$

Here and below $r_0 = r_e$ and $r_1 = r_o$. The terms with $j = 0$ in expressions (2.41) and (2.42) correspond to the even part of the field and the terms with $j = 1$—to the odd. For axial incidence, when $\alpha = 0$, these expressions have ambiguities of the type $0/0$. Computing the limit as $\alpha \to 0$ (see formulas 14.1.4–7 in [1]), it is easy to conclude that the odd part of the field vanishes and the even part in view of the limit

$$\lim_{\alpha \to 0} \frac{F_{-3/4}\left(t, \frac{\alpha^2}{2}\right)}{\sqrt{\alpha}} = \frac{\mathrm{e}^{-\pi t/2}}{2\sqrt{\pi}} \sqrt{\Gamma\left(\frac{1}{4} + \mathrm{i}t\right) \Gamma\left(\frac{1}{4} - \mathrm{i}t\right)}$$

remains finite.

2.5.3 Reduction to Fock Asymptotics in the Case of $\chi \to \infty$

The above-presented formulas present the leading order asymptotic approximation for the field of diffraction by an elongated ellipse. The rate of elongation is characterized by parameter χ. When χ increases, the body becomes less and less elongated. Therefore for $\chi \to \infty$, the asymptotic formulas for the currents of the forward wave process should reduce to the usual Fock asymptotics (1.9) and (1.10). We check that considering the case of plane wave diffraction and the small vicinity of the light-shadow boundary.

We consider the currents corresponding to the forward-going wave process, which are expressed by formulas (2.41) and (2.42) in which only the terms with the factor $e^{ikp\eta}$ should be retained. We present derivations only for the TM case, while formula (2.41) can be analyzed in a similar way. When χ is large, the main contribution to the integral in (2.42) is due to the vicinity of point $t = -\chi/4$. On this hand, it is convenient to change the integration variable to $-t$. So, the induced current of the forward wave is expressed by formula

$$J_{TM}^{\text{forward}} \sim \frac{4\sqrt{2}\mathrm{e}^{\mathrm{i}kp\eta}\,\chi^{3/4}}{\sqrt[4]{1-\eta^2}\sqrt{\pi\alpha}} \int\limits_{-\infty}^{+\infty} \left(\frac{1+\eta}{1-\eta}\right)^{\mathrm{i}t} \times$$
$$\times \sum_{j=0,1} \frac{(\mathrm{i}\,\mathrm{sign}(\vartheta_0 x))^j \; F_{-\frac{3-2j}{4}}\left(-t,\frac{\alpha^2}{2}\right)}{H^{+}_{\frac{2j-3}{4}}\left(-t,\frac{\chi}{2}\right) - 2\chi \dot{H}^{+}_{\frac{2j-3}{4}}\left(t,\frac{\chi}{2}\right)} \mathrm{d}t. \quad (2.43)$$

We substitute the asymptotic formula (A.26) into (2.43) and consider η to be small, such that

$$\left(\frac{1+\eta}{1-\eta}\right)^{\mathrm{i}t} \approx \mathrm{e}^{2\mathrm{i}t\eta}.$$

Thus, we get

$$J_{TM}^{\text{forward}} \sim -\frac{2^{5/6}\mathrm{e}^{\mathrm{i}kp\eta}}{\sqrt{\pi}}\chi^{-1/4} \int \frac{\mathrm{e}^{2\mathrm{i}\eta t + 2\mathrm{i}\sqrt{t}\alpha\sigma}}{\dot{w}_1(\xi)} t^{1/6}\mathrm{d}t,$$

where

$$\xi = \left(2t - \frac{\chi}{2}\right)(2t)^{-1/3}.$$

In this integral, we change the integration variable to ξ, assuming

$$t \approx \frac{\chi}{4} + \frac{\xi}{2}\left(\frac{\chi}{2}\right)^{1/3}.$$

At the leading order, we get

$$J_{TM}^{\text{forward}} \sim \frac{1}{\sqrt{\pi}} \mathrm{e}^{\mathrm{i}kb\eta + \mathrm{i}\sqrt{\chi}\alpha} \int \frac{\mathrm{e}^{\mathrm{i}\sigma\xi}}{\dot{w}_1(\xi)} \mathrm{d}\xi,$$

where

$$\sigma = \eta\left(\frac{\chi}{2}\right)^{1/3} + 2^{-1/3}\chi^{-1/6}\alpha.$$

Taking into account the expressions for α and χ, we find

$$kb\eta + \sqrt{\chi}\alpha = kz + ka\vartheta_0, \qquad \sigma = \left(\frac{k\rho}{2}\right)^{1/3}\left[\frac{z}{\rho_0} + \vartheta_0\right],$$

where $\rho_0 = b^2/a$ is the radius of curvature of the surface at $\eta = 0$.

Thus, for large χ the asymptotic approximation (2.43) reduces to the classical Fock asymptotic approximation for the current induced by the TM incident wave. Similarly, it can be shown that asymptotic representation (2.42) in its part corresponding to the forward wave reduces to the classical Fock asymptotic formula for the induced current in the case of TE polarization.

2.5.4 The Case of Line Source Field Incidence

In view of the noted relation between the amplitude functions $A_{o,e}(t)$ for the plane wave incidence and in the case of a line source field incidence, which is specified at the end of Sect. 2.3.3, it is a simple matter to write the formulas for the induced currents. For that, we replace α^2 in (2.41) and (2.42) with Υ_0, insert factor $((\zeta_0 - 1)/(\zeta_0 + 1))^{it}$ below the sign of integration and include external multiplier Ω given by (2.32).

2.5.5 Accuracy of Approximation and Test Examples

Here, we present some test examples that help us to analyze the applicability conditions for the asymptotic approximations (2.41) and (2.42). We compare the currents computed numerically with their asymptotic approximations.

Figures 2.2 and 2.3 present the currents on the surface of two elliptic cylinders, one with $b = 0.2$m, $a = 0.05$m, the other with $b = 0.2$m, $a = 0.02$m, induced by plane waves with $k = 10$, 20, 50 and 100 incident along the major axis of the ellipse. We compare asymptotic approximation with numerical results, obtained with the pdetool package of Matlab, and can conclude that the asymptotic formulas give sufficiently good approximation already for $kb = 2$. For $kb = 10$ the agreement is the best. For larger kb, Matlab fails to perform correct computations. We can note also that for a more elongated ellipse in Fig. 2.3, the agreement is a little better, which is natural, because the asymptotic formulas are derived under the assumption that the ellipses are strongly elongated.

It is worth noting that the classical Fock asymptotics gives overestimated values for the induced current. Indeed, on the light-shadow boundary it predicts $J \approx 1.399$ in the TM case, while values on the left-hand sides of Figs. 2.2 and 2.3 are much smaller.

For the tests with the incidence at an angle to the major axis, we consider only the incidence of the TM plane wave with $k = 50$. The results, shown in Figs. 2.4 and 2.5, allow us to conclude that in the case of a more elongated cross-section one can use the

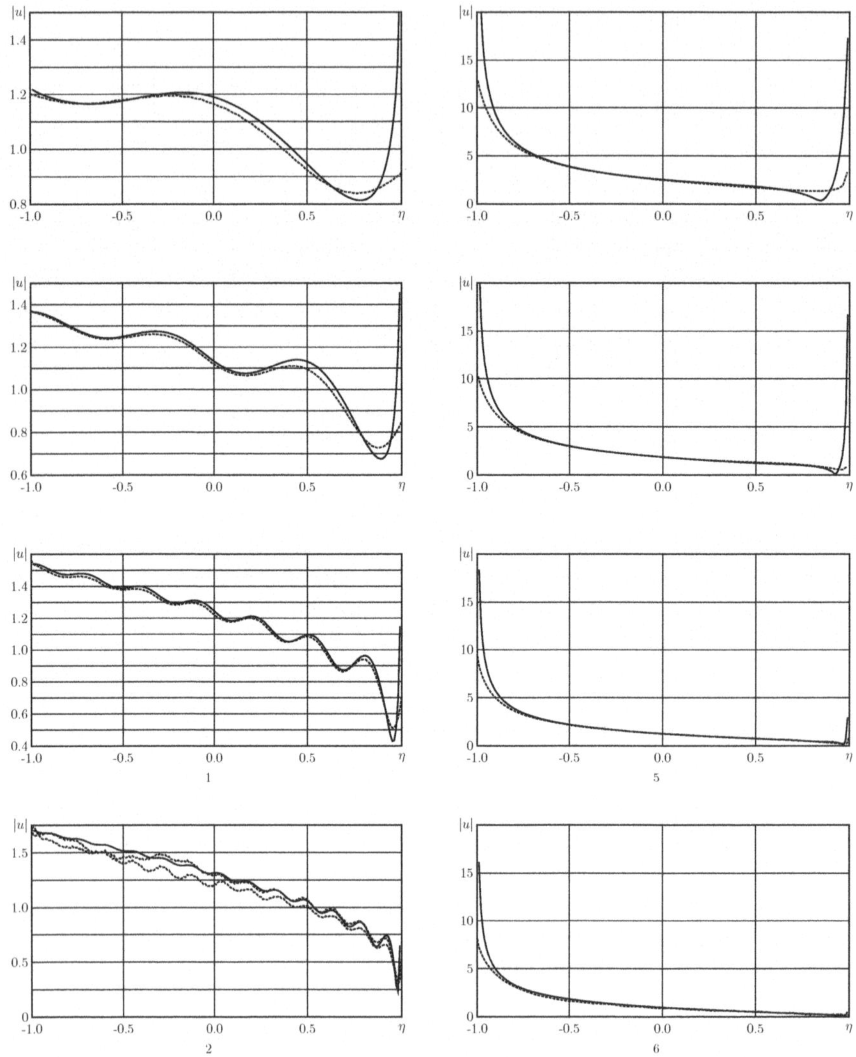

Fig. 2.2 TM (left) and TE (right) currents on the elliptic cylinder with $a = 0.05$, $b = 0.2$ for $k = 10, 20, 50$ and 100 (downwards). Solid—asymptotics; dashed—numerical

asymptotic approximation (2.42) for a larger interval of the incidence angles. For $a : b = 1 : 10$, noticeable errors appear only for $\vartheta = 20°$, while for less elongated ellipse with $a : b = 1 : 4$ we can see some differences already for $\vartheta_0 = 15°$. Certainly, the asymptotic parameter $kb = 10$ is not sufficiently large. Increasing the frequency should improve the accuracy of the asymptotic representation. However, it is worth noting that the increase of kp causes the decrease of the interval of angles ϑ_0, in order to maintain $\alpha = \sqrt{kp}\,\vartheta$ not large.

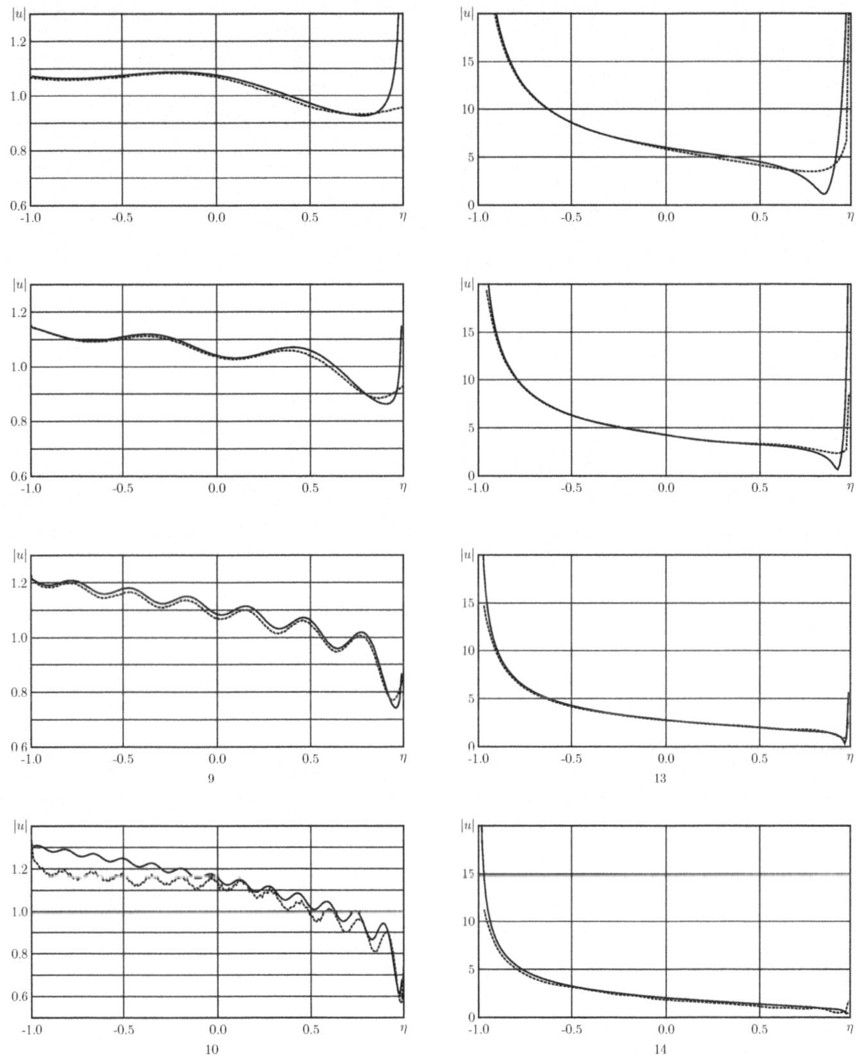

Fig. 2.3 TM (left) and TE (right) currents on the elliptic cylinder with $a = 0.02$, $b = 0.2$ for $k = 10, 20, 50$ and 100 (downwards). Solid—asymptotics; dashed—numerical

2.5.6 *Results for the Line Source Field Diffraction*

To illustrate the line source diffraction, we present in Fig. 2.6 the density plots showing the amplitude of the TM total field distribution in the boundary layer near the surface. We use the following approximations. In the boundary layer, except the small vicinities of the tips, the total field is approximated by the sum $u^{\text{inc}} + u^{\text{forward}} + u^{\text{back}}$, where forward wave u^{forward} is expressed by (2.14) where we substitute reflection

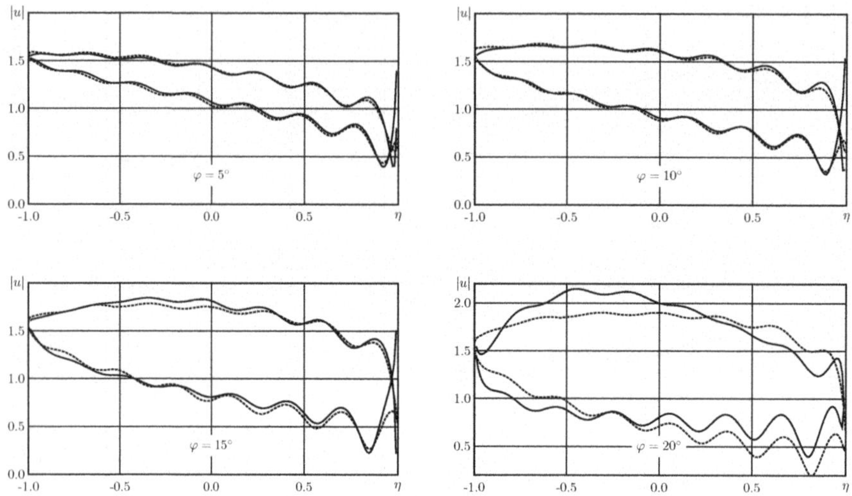

Fig. 2.4 Diffraction by the ellipse with $a = 0.05$, $b = 0.2$ at $kb = 10$ and different angles of incidence. Solid—asymptotics; dashed—numerical

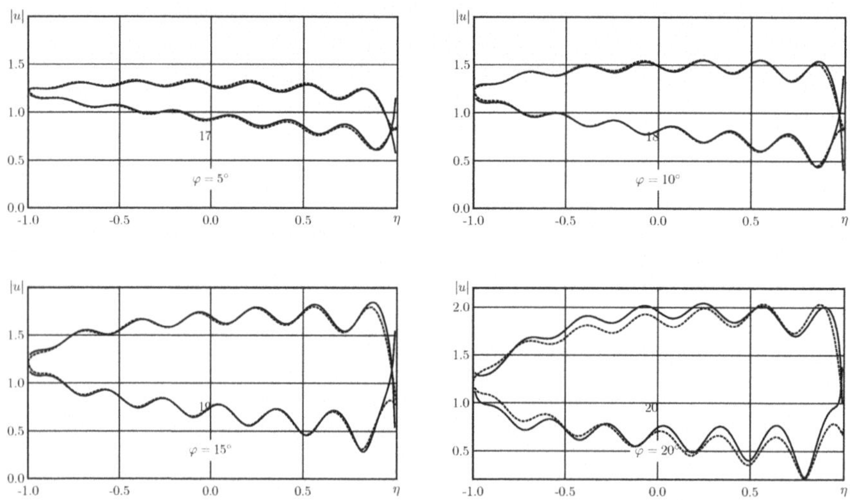

Fig. 2.5 Diffraction by the ellipse with $a = 0.02$, $b = 0.2$ at $kb = 10$ and different angles of incidence. Solid—asymptotics; dashed—numerical

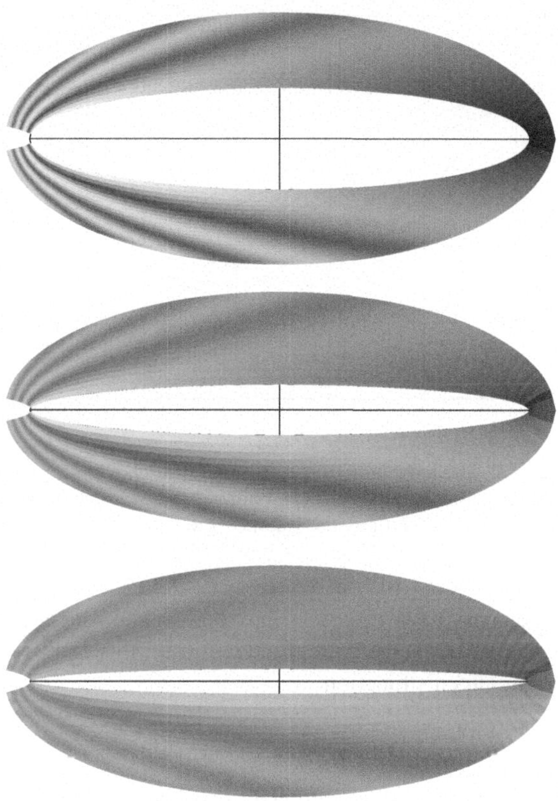

Fig. 2.6 The distribution of the acoustic pressure of the point source, located at $z_0 = -5b$, $x_0 = -b$, near the surface of the rigid elliptic cylinders with $kb = 100$ and different ka: $ka = 20$, $ka = 10$ and $ka = 5$. The colors from black to white correspond to the field amplitudes from 0 to 2

coefficients $R_{o,e}$ given by (2.16) and the amplitudes $A_{o,e}$ given by (2.20) and (2.22). The backward wave u^{back} is expressed by (2.33) with coefficients $r_{o,e}$ substituted there from (2.37). In the vicinity of the rear tip of the ellipse, we use the representation (2.34) in which we substitute $P_{o,e}$ from (2.36). The amplitude of the source is chosen such that the incident field amplitude at the center of the ellipse is equal to one.

The influence of the point source location on the field distribution is illustrated in Fig. 2.7, where we also consider the case of TM polarization.

2.6 The Far Field

2.6.1 Far Field in the Forward Cone

The previous section allows concluding that the asymptotic formulas (2.42) and (2.41) provide accurate approximations for the currents on the surface of a perfectly conducting elliptic cylinder for a wide range of parameters. Substituting these

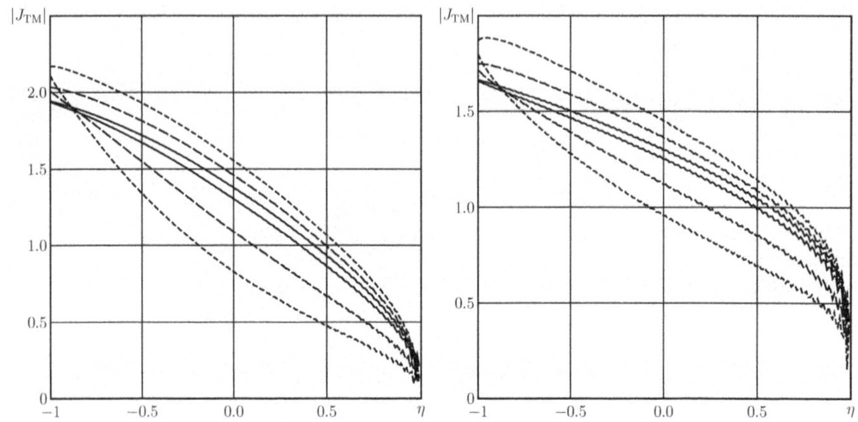

Fig. 2.7 The amplitude of the TM total field on elliptic cylinder with $kb = 100$, $ka = 20$ (left) and $kb = 100$, $ka = 10$ (right) for different positions of the line source: $x_0 = -2b$ and $z_0 = -50b$ (solid), $z_0 = -10b$ (long-dashed) and $z_0 = -5b$ (short-dashed curves)

currents as passive sources into the Kirchhoff integral enables us to find the asymptotics of the far field amplitude. Toward this end, we use the representation of the scalar Green's function

$$G(\mathbf{r}, \mathbf{r}') = \frac{i}{4} H_0^{(1)}(k|\mathbf{r} - \mathbf{r}'|) \tag{2.44}$$

with $\mathbf{r}'$ located in the boundary layer near the surface and $\mathbf{r}$ tending to infinity. Let us define the far field amplitude Ψ by using the asymptotic formula [15]

$$u(\mathbf{r}) \sim \frac{e^{ikr}}{\sqrt{kr}} \Psi(\vartheta), \qquad |\mathbf{r}| = r \to +\infty.$$

Then the far field amplitude of Green's function (2.44) can be written as

$$\Psi_G = \frac{e^{i\pi/4}}{2\sqrt{2\pi}} \exp(-ikz_0 \cos\vartheta - ikx_0 \sin\vartheta).$$

Consequently, as a function of the point of the source, it coincides, within the multiplicative factor, with a plane wave incident from the direction opposite to that along which $\mathbf{r}$ tends to infinity. The above observation enables us to write the asymptotic representation for Ψ_G by changing η to $-\eta$ in (2.7) and (2.14). Thus, we have

$$\Psi_G(\vartheta,\eta,\tau) = \frac{e^{-ikp\eta+i\pi/4}}{4\pi^2\sqrt[4]{1-\eta^2}\sqrt[4]{\chi\tau}\sqrt{\beta}}\int_{-\infty}^{+\infty}\left(\frac{1-\eta}{1+\eta}\right)^{-is}\sum_{\ell=0,1}(4i\,\mathrm{sign}(\beta))^{\ell}\times$$
$$\times\, M_{is,\frac{2\ell-1}{4}}(-i\chi\tau)M_{is,\frac{2\ell-1}{4}}(i\beta^2)\Gamma\left(\tfrac{2\ell+1}{4}+is\right)\Gamma\left(\tfrac{2\ell+1}{4}-is\right)ds. \quad (2.45)$$

Here, $\beta=\sqrt{kp}\vartheta$ and it is assumed to be bounded when $kp\to\infty$. For the convenience of further derivations compared to (2.14), we replaced j with ℓ and the integration variable t with s. By using the Coulomb wave functions in (2.45), we can write

$$\Psi_G(\vartheta,\eta,\tau) = \frac{e^{i\pi/4}}{\pi}\frac{e^{-ikp\eta}}{\sqrt[4]{1-\eta^2}\sqrt[4]{\chi\tau}\sqrt{\beta}}\times$$
$$\times\int_{-\infty}^{+\infty}\left(\frac{1-\eta}{1+\eta}\right)^{-is}\sum_{\ell=0,1}(i\sigma)^{\ell}F_{-\frac{3-2\ell}{4}}\left(s,\frac{\beta^2}{2}\right)F_{-\frac{3-2\ell}{4}}\left(-s,\frac{\chi\tau}{2}\right)ds. \quad (2.46)$$

In the TM case, the far field amplitude of the scattered field can be expressed by using the formula

$$\Psi_{TM} = \int\frac{\partial\Psi_G}{\partial n}J_{TM}dS, \quad (2.47)$$

where the integration is performed along the ellipse, S denotes the arc-length and n is the external normal to the surface.

If the direction of observation is close to the major axis, the main contribution to the integral in (2.47) comes from the integration over the middle part of the surface, where quick oscillations of Ψ_G and of J_{TM} compensate each other. In this domain, we can use the asymptotics (2.42) and (2.46) and substituting these expressions in (2.47). Furthermore, in the middle part of the strongly elongated ellipse we have $dS\approx b\,d\eta$ and

$$\frac{\partial}{\partial n}\approx\frac{2}{a}\frac{1}{\sqrt{1-\eta^2}}\frac{\partial}{\partial\tau}.$$

When combining the integrals over the upper and lower halves of the ellipse, the terms with $\ell\neq j$ compensate each other and the other terms double. Finally, we change the order of integration and compute the integral with respect to η first. Noting that (see the integral transform in the Appendix A.4)

$$\int_{-1}^{1}\left(\frac{1-\eta}{1+\eta}\right)^{i(t-s)}\frac{d\eta}{\sqrt{1-\eta^2}} = \pi\delta(t-s),$$

we obtain the leading order approximation for the far field amplitude in the following form:

$$\Psi_{TM} = -\frac{8e^{i\pi/4}}{\sqrt{2\pi}}\sqrt{\frac{kb}{\alpha\beta}}\int\limits_{-\infty}^{+\infty}\left\{F_{-\frac{3}{4}}\left(t,\frac{\alpha^2}{2}\right)F_{-\frac{3}{4}}\left(t,\frac{\beta^2}{2}\right)\times\right.$$
$$\times\frac{F_{-\frac{3}{4}}\left(-t,\frac{\chi}{2}\right)-2\chi\dot{F}_{-\frac{3}{4}}\left(-t,\frac{\chi}{2}\right)}{H^{+}_{-\frac{3}{4}}\left(-t,\frac{\chi}{2}\right)-2\chi\dot{H}^{+}_{-\frac{3}{4}}\left(-t,\frac{\chi}{2}\right)}$$
$$+\operatorname{sign}(\alpha\beta)F_{-\frac{1}{4}}\left(t,\frac{\alpha^2}{2}\right)F_{-\frac{1}{4}}\left(t,\frac{\beta^2}{2}\right)\times$$
$$\left.\times\frac{F_{-\frac{1}{4}}\left(-t,\frac{\chi}{2}\right)-2\chi\dot{F}_{-\frac{1}{4}}\left(-t,\frac{\chi}{2}\right)}{H^{+}_{-\frac{1}{4}}\left(-t,\frac{\chi}{2}\right)-2\chi\dot{H}^{+}_{-\frac{1}{4}}\left(-t,\frac{\chi}{2}\right)}\right\}dt. \tag{2.48}$$

In the TE case, we can express the far field amplitude as

$$\Psi_{TE} = -k\int \Psi_G J_{TE}\, dS.$$

Substituting the asymptotics (2.41) and (2.46) in this formula, and performing similar derivations as above, we get

$$\Psi_{TE} = -\frac{8e^{i\pi/4}}{\sqrt{2\pi}}\sqrt{\frac{kb}{\alpha\beta}}\int\limits_{-\infty}^{+\infty}\left\{F_{-\frac{3}{4}}\left(t,\frac{\alpha^2}{2}\right)F_{-\frac{3}{4}}\left(t,\frac{\beta^2}{2}\right)\frac{F_{-\frac{3}{4}}\left(-t,\frac{\chi}{2}\right)}{H^{+}_{-\frac{3}{4}}\left(-t,\frac{\chi}{2}\right)}\right.$$
$$\left.+\operatorname{sign}(\alpha\beta)F_{-\frac{1}{4}}\left(t,\frac{\alpha^2}{2}\right)F_{-\frac{1}{4}}\left(t,\frac{\beta^2}{2}\right)\frac{F_{-\frac{1}{4}}\left(-t,\frac{\chi}{2}\right)}{H^{+}_{-\frac{1}{4}}\left(-t,\frac{\chi}{2}\right)}\right\}dt. \tag{2.49}$$

Expressions (2.48) and (2.49) show that at the leading-order, the far field amplitude is determined by three parameters: the parameter $\chi = ka'b$, which characterizes the rate of elongation; and two scaled angles: the angle of incidence $\alpha = \sqrt{kb}\vartheta_0$ and the angle of observation $\beta = \sqrt{kb}\vartheta$.

Below, we present the asymptotic results computed by using expressions (2.48) and (2.49). We consider elliptic cylinders of fixed wavelength, such that $kb = 100$, but with different rates of elongation, starting from the limiting case of the strip, for which $\chi = 0$, and ending with the case of the circular cylinder, for which $\chi = kb$.

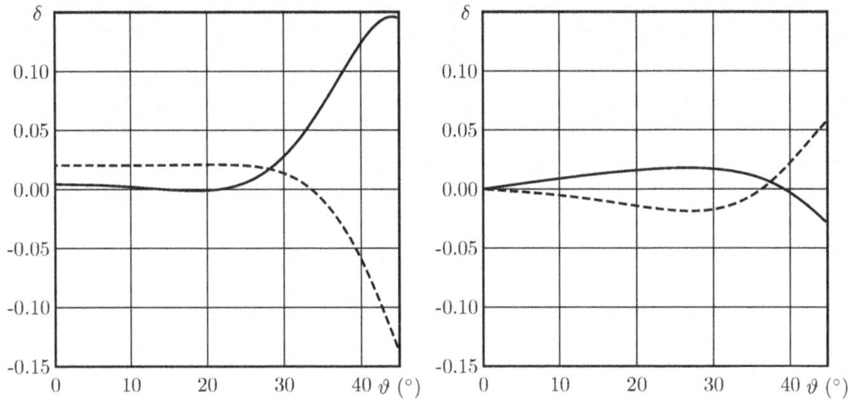

Fig. 2.8 The error $\delta = \Psi_{\text{asympt.}} - \Psi_{\text{numeric}}$ for the diffraction by the strip with $kb = 10, \vartheta_0 = 5°$. Solid curve corresponds to the real part, dashed—to the imaginary part of δ

2.6.2 Numerical Examples and Tests for the Far Field

It is worth noting that the case of the circular cylinder cannot be considered in the elliptic coordinates which become singular in this case as the focal distance $2p$ becomes equal to zero. However, (2.48) and (2.49) do not contain ambiguities in this case. For the case of the circular cylinder, the elongation parameter χ is equal to the asymptotic parameter kb, i.e. it is asymptotically large, but not $O(1)$ as was assumed to be the case when deriving the asymptotic expansion. However, such large values seem to be possible, because with $\chi \to \infty$ the formulas for the currents reduce to the classical results of Fock that are evidently valid for the case of the circular cylinder.

For the case of the strip, we compare our asymptotic results with the far field amplitude computed with the boundary integral method.[1] Figure 2.8 presents the errors for the test example corresponding to $kb = 10$ and $\vartheta_0 = 5°$. The error is less than 1%. This error can appear because the asymptotic parameter kb is not sufficiently large. One can note that the error increases with the increase of the angle. This is also natural in view of our assumption that the angles ϑ and ϑ_0 are small.

Figures 2.9 and 2.10 present the Radar Cross-Section (RCS=$2\pi|\Psi|^2$) for the plane wave incident along the major axis, and for an incidence at angle $\vartheta_0 = 5°$, for which $\beta \approx 0.276$. Figure 2.9 deals with TE polarization. For the case of the strip, the RCS plots are symmetric which agrees with the symmetry of the problem. For cylinders with $ka > 0$, this symmetry is preserved only for the case of axial incidence. The maximum level of the RCS is located approximately along the direction of incidence. For thicker cylinders, i.e. for $ka = 20$ and $ka = 40$, the plots for the axial incidence case and those with an angle to the axis differ only by the 5° shift, and this is an exact characteristic of the RCS for the case of the circular cylinder. For the case of

[1] Computations were performed by A. V. Korol'kov (Dept. of Physics, Moscow State University, Russia.).

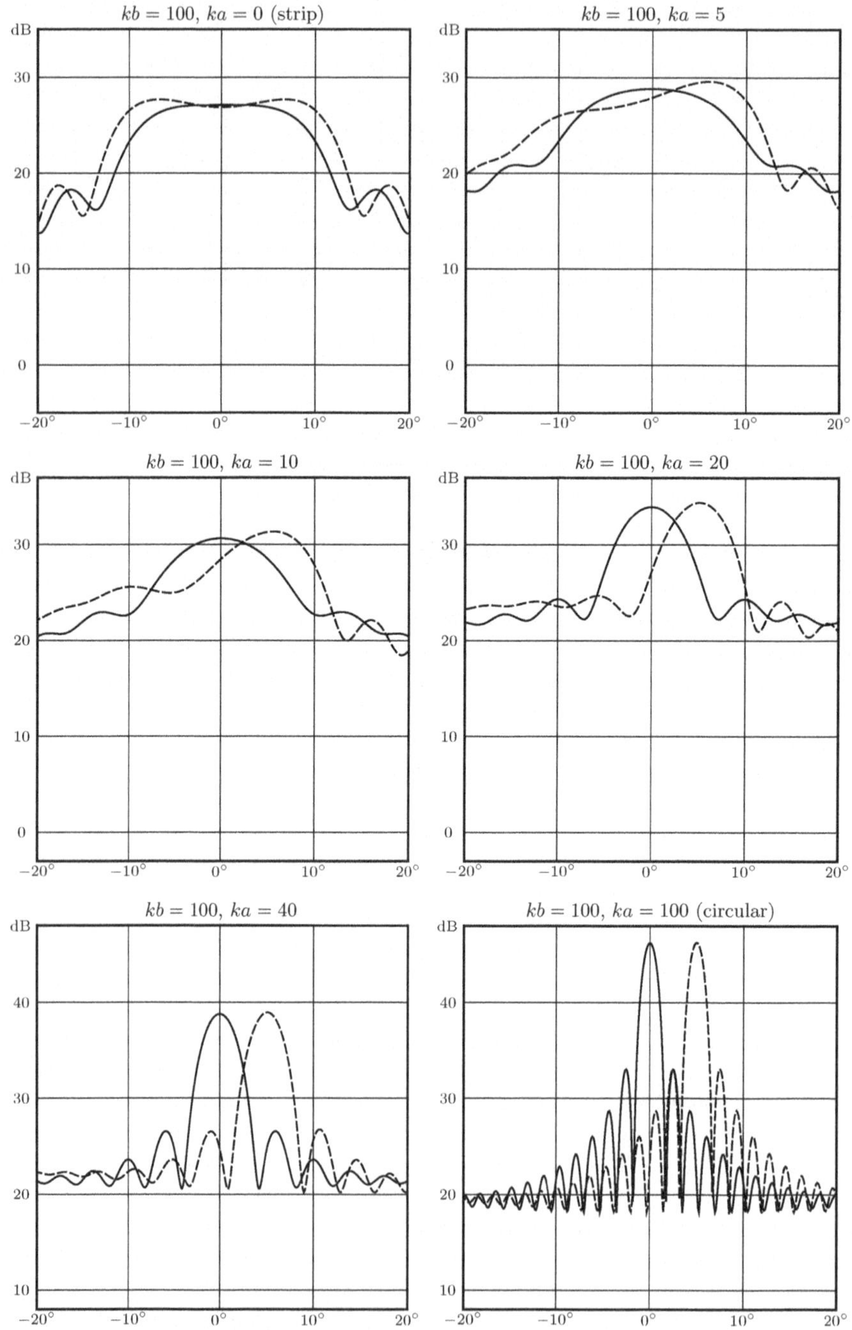

Fig. 2.9 RCS of plane TE wave incident on elliptic cylinders with $kb = 100$ and different aspect ratios a/b. The case of incidence along the major axis (solid) and the case of incidence at the angle of 5° (dashed)

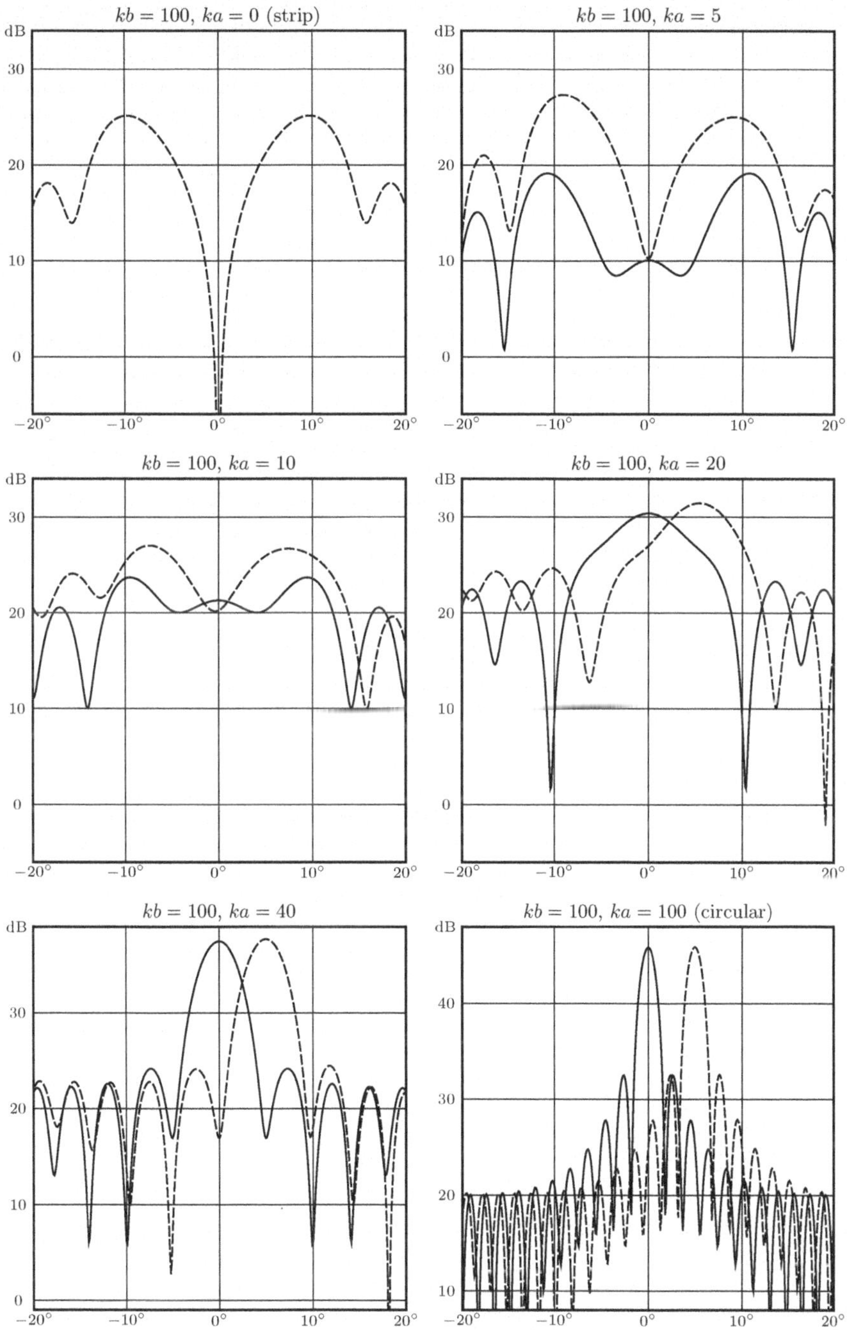

Fig. 2.10 RCS of plane TM wave incident on elliptic cylinders with $kb = 100$ and different aspect ratios a/b. The case of incidence along the major axis (solid) and the case of incidence at the angle of 5° (dashed)

TM polarization, presented in Fig. 2.10, we see that the main beam of the RCS may be directed not along the direction of incidence, but along a mirror direction instead. See the case of $ka = 5$. For larger ka, starting with $ka = 40$, we see that the RCS for axial incidence and the incidence at 5° differs by the shift, in the same way as it does for the TE case.

2.6.3 Far Field in the Backward Cone

The developed asymptotic approach allows the far fields to be analyzed also in the backward direction. The back-scattered field is the sum of two contributions, the reflection from the specular point and the radiation of the backward wave. Since we assume that the directions of incidence and observation are close to the axial, the specular point is in a small vicinity of the tip of the ellipse. In this domain, we can approximate the surface with the surface of a parabolic cylinder. Diffraction by a parabolic cylinder is a classical problem [12] which can be tackled by using the parabolic coordinates

$$z + p = \frac{v - w}{2k}, \qquad x = \frac{\sqrt{vw}}{k}.$$

In these coordinates, the surface is specified by the equation $w = \chi$.

For the incident plane wave

$$u^{\text{inc}} = \exp\left(\mathrm{i}kz\cos\vartheta_0 + \mathrm{i}kx\sin\vartheta_0\right),$$

it is convenient to use the representation [13, 16]

$$u^{\text{inc}} = \frac{\mathrm{e}^{-\mathrm{i}kp\cos\vartheta_0}}{\sqrt[4]{vw}\cos(\vartheta_0/2)} \sum_{j=0}^{+\infty} \frac{(\mathrm{i}\tan(\vartheta_0/2))^j}{j!} W_{\frac{2j+1}{4},-\frac{1}{4}}(-\mathrm{i}v) W_{\frac{2j+1}{4},-\frac{1}{4}}(\mathrm{i}w).$$

Then the reflected field can be represented as

$$u = \frac{\mathrm{e}^{-\mathrm{i}kp\cos\vartheta_0}}{\sqrt[4]{vw}\cos(\vartheta_0/2)} \sum_{j=0}^{+\infty} \frac{(\mathrm{i}\tan(\vartheta_0/2))^j}{j!} W_{\frac{2j+1}{4},-\frac{1}{4}}(-\mathrm{i}v) R_j W_{-\frac{2j+1}{4},-\frac{1}{4}}(-\mathrm{i}w), \tag{2.50}$$

where the reflection coefficients R_j are defined from the boundary conditions. For the TE and TM cases we find, respectively,

$$R_j^{TE} = -\frac{W_{\frac{2j+1}{4},-\frac{1}{4}}(\mathrm{i}\chi)}{W_{-\frac{2j+1}{4},-\frac{1}{4}}(-\mathrm{i}\chi)},$$

$$R_j^{TM} = -\frac{W_{\frac{2j+1}{4},-\frac{1}{4}}(\mathrm{i}\chi) - 4\mathrm{i}\chi \dot{W}_{\frac{2j+1}{4},-\frac{1}{4}}(\mathrm{i}\chi)}{W_{-\frac{2j+1}{4},-\frac{1}{4}}(-\mathrm{i}\chi) + 4\mathrm{i}\chi \dot{W}_{-\frac{2j+1}{4},-\frac{1}{4}}(-\mathrm{i}\chi)}.$$

In the limiting case of the strip, i.e. for $\chi = 0$, the Whittaker functions in the above formulas have singularities. The limits of the reflection coefficients are

$$R_j^{TE} = -\mathrm{e}^{\mathrm{i}\pi/4}\frac{\Gamma\left(1+\frac{j}{2}\right)}{\Gamma\left(\frac{1-j}{2}\right)} \quad \text{and} \quad R_j^{TM} = \mathrm{e}^{-\mathrm{i}\pi/4}\frac{\Gamma\left(\frac{j+1}{2}\right)}{\Gamma\left(-\frac{j}{2}\right)}.$$

By letting the observation point to infinity, we can replace the Whittaker functions with their asymptotics (A.23)

$$W_{\lambda,\ell}(\zeta) \sim \mathrm{e}^{-\zeta/2}\zeta^{\lambda}, \qquad |\zeta| \to +\infty.$$

The outgoing wave is formed by the second terms in the braces in (2.50), and we can write

$$u^{\text{scat}} = \frac{\mathrm{e}^{\mathrm{i}\frac{v-w}{2} - \mathrm{i}kp\cos\vartheta_0}}{\sqrt[4]{vw}} \frac{1}{\cos(\vartheta_0/2)} \sum_{j=0}^{+\infty} \frac{R_j}{j!} (2\mathrm{i}\tan(\vartheta_0/2))^j \left(\frac{v}{w}\right)^{\frac{2j+1}{4}}.$$

Accounting for $v = k\sqrt{(z-p)^2 + x^2} + k(z+p)$, $w = k\sqrt{(z-p)^2 + x^2} - k(z+p)$, the above expression for the scattered field takes the form

$$u^{\text{scat}} \sim \frac{\mathrm{e}^{\mathrm{i}kr}}{\sqrt{kr}} \frac{\mathrm{e}^{-\mathrm{i}kp(\cos\vartheta_0 + \cos\vartheta')}}{\sqrt{2}\cos(\vartheta_0/2)\cos(\vartheta'/2)} \sum_{j=0}^{+\infty} \frac{R_j}{j!} \left(2\mathrm{i}\tan\left(\frac{\vartheta_0}{2}\right)\tan\left(\frac{\vartheta'}{2}\right)\right)^j, \tag{2.51}$$

where $\vartheta' = \pi + \vartheta$.

The backward wave radiation can be considered in the same manner as that for the forward wave. The amplitude of the backward wave differs from the amplitude of the forward wave by the multipliers r_j that are the reflection coefficients of the even and odd parts of the wave from the shaded tip of the elliptic cylinder. These coefficients are given in (2.37). Therefore, the expressions for the far field amplitudes of the backward wave radiation differ from those given by formulas (2.48) and (2.49) by the additional multipliers r_j, which should be included in the integral, and we also should replace β with $\beta' = \sqrt{kb}\vartheta'$, which accounts for the change in the direction of observation.

Computations show that for the TE case, the contribution of the backward wave is negligible. For the strip of length $kb = 100$, it is approximately 70 dB lower than the contribution of the reflection from the specular point, and for the more thick elliptic cylinders this difference becomes larger. This effect is attributable to the rapid attenuation of TE creeping waves. Thus, the RCS in the backward direction is the same as in the case of the parabolic cylinder, and it is almost independent of the incident and observation angles.

For the TM case, the creeping waves are less attenuated and the contribution of the backward field is larger. It cannot be neglected if the rate of elongation of the elliptic cylinder is high. Figure 2.11 presents the RCS in the backward directions for the TM wave, for the axial incidence case as well as for an angle of 5° from the major axis. For the case of the strip (upper left plots), we see that the RCS is nonzero only when neither the observation nor the incidence angles are zero. When the cylinder is not absolutely flat, the RCS quickly increases with the elongation parameter χ, so it is already 35 dB higher than that for the strip for $\chi = 10^{-4}$ (the upper right plots). For the axial incidence case, the contributions of the specular reflection and the backward wave are of the same order of magnitude. However, for the incidence at an angle to the axis, the main contribution arises from the backward wave radiation, while the specular reflection remains small. If we further increase the elongation parameter χ, the magnitude of the specular reflection increases while the magnitude of the backward wave remains approximately the same. So, for the cylinders with aspect ratios less than 1:20, the main effect of backscattering is due to the specular reflection, and the polarization of the incident wave is not so significant for this case.

We should remark that in contrast to the forward scattering case, the asymptotic formulas for the back-scattered wave are not applicable for scattering by thick cylinders. On the one hand, the approximation of such elliptic cylinders with parabolic cylinders is not accurate, and the expression given in (2.51) for the contribution of specular reflection cannot be used. In addition, the method for finding the reflection coefficients r_j is also invalid in this case.

2.7 Conclusion

We considered in this chapter the 2D problem of diffraction. The derivations demonstrate the main peculiarities of the parabolic equation method modified to the problems of diffraction by strongly elongated surfaces. The procedure may be thought of as consisting of the following steps. First, the approximation in the vicinity of the surface is constructed. The field is represented as the sum of the forward and backward wave processes. Both are treated in the parabolic equation approximation, but with different (opposite) quick factors. The leading order term approximations for the fields are expressed in the form of the integrals which contain solutions of the Whittaker equation. In these integral representations, one can identify specific factors each of which is responsible for the particular characteristics of the problem. One of such factors is the amplitude factor $A_{e,o}$ of the incident wave. We computed these factors for the case of plane wave incidence and for the case of the cylindrical incident wave originating from a line source. The other factor is the reflection coefficient $R_{o,e}(t)$ by which the integral representations in the TE and TM cases differ from each other. The reflection coefficients are given by (2.15) and (2.16) correspondingly. The backward wave representation contains the additional multiplier $r_{o,e}$ which may

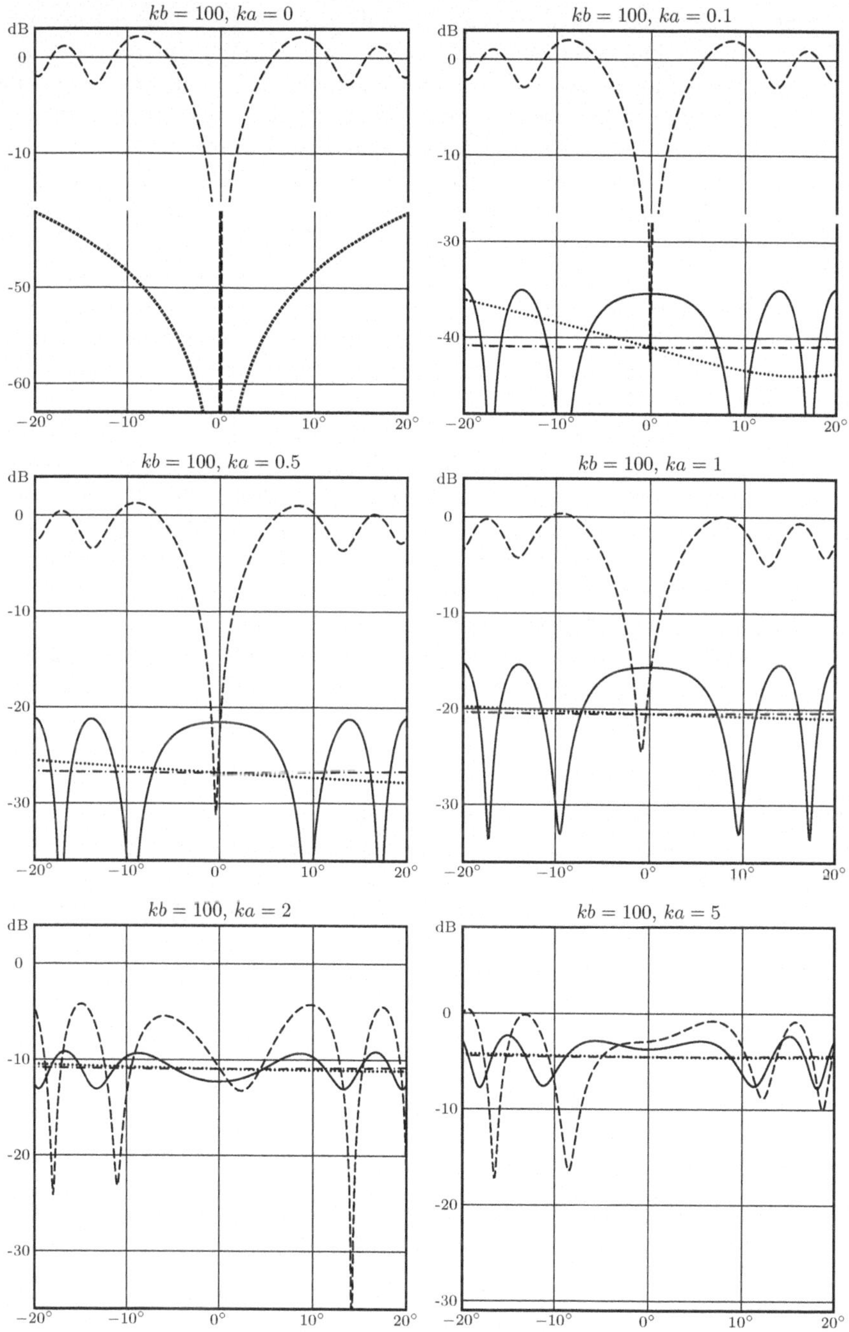

Fig. 2.11 The RCS in backward directions for the case of TM incident wave axially (solid) and at $\vartheta_0 = 5°$ to the axis (dashed). The contribution of the specular reflection is plotted for $\vartheta_0 = 0$ in dash-dotted and for $\vartheta_0 = 5°$ in dotted lines

be thought of as the reflection coefficient from the rear tip of the ellipse. This coefficient is influenced by the effect of the forward wave encircling the rear end of the body. We have found expressions (2.37) for these multipliers in the case of smooth rear extremity of the surface which we approximated with the appropriate parabolic surface. Though we have not considered the other shapes of the rear ending of the body, it is natural to expect that for example in the case of sharp ending the integral representations will differ only by the replacement of factors $r_{o,e}$ with other ones that can be obtained by matching the corresponding local approximations.

In the second step of the procedure, the far field approximation is obtained by applying Green's formula with the representations for the induced currents substituted into it. We have derived these approximations using the currents in the middle part of the surface, which enables us to get the far field approximations in the paraxial forward and backward directions.

All the asymptotic expressions involve the elongation parameter χ, which crucially influences the fields. In the case of $\chi \to 0$, the surface shrinks and the elliptic cylinder transforms into the strip. The approximations derived in this chapter remain valid in this case. A more detailed analysis of the fields in the problem of diffraction by the strip can be found in [5, 7] and [10]. When parameter χ increases, which corresponds to less and less elongated bodies, the approximations reduce to the classical Fock asymptotics, which we checked in Sect. 2.5.3. Thus, we can conclude that the approximations derived in this chapter are uniform with respect to the rate of elongation and can be used in the problems of diffraction by an ordinary body and by the strongly elongated body up to its limiting case of the strip.

To conclude this chapter, we point out the following generalizations. The case of acoustic diffraction of a point source field by an elliptic cylinder with a strongly elongated cross-section is considered in [6]. In this case, the field depends on y coordinate and this makes the problem more complicated. However, the effects seem to preserve locality with respect to the y coordinate and the leading order approximation appears the same as in the case of plane wave incidence at an appropriate angle. The other extension of the technique developed in this chapter is presented in [22], which considers truncated elliptic cylinders in the frame of the Physical theory of diffraction [24].

References

1. M. Abramowitz, I. A. Stegun (ed.), *Handbook of Mathematical Functions with Formulas, Graphs and Mathematical Tables* (National Bureau of Standards, Appl. Math. Series 55, 1964)
2. I.V. Andronov, Diffraction at an elliptical cylinder with a strongly prolate cross section. Acoust. Phys. **60**(3), 237–244 (2014)
3. I.V. Andronov, High-frequency diffraction by an elliptic cylinder: the near field. J. Electromagn. Waves Appl. **28**(18), 2318–2326 (2014)
4. I.V. Andronov, Uniform asymptotics for high-frequency diffraction by an elliptic cylinder. Antennas Wirel. Propag. Lett. **14**, 1204–1206 (2015)
5. I.V. Andronov, On high-Frequency scattering by a strip at nearly grazing incidence. Acoust. Phys. **62**(4), 399–404 (2016)

6. I.V. Andronov, High-frequency diffraction of a point source field by a strongly elongated elliptic cylinder. Acta Acustica United Acustica **105**(6), 912–917 (2019)
7. I.V. Andronov, D. Bouche, Diffraction by a strip at almost grazing angle. J. Sound Vib. **374**, 185–198 (2016)
8. I.V. Andronov, Yu.A. Lavrov, Scattering by an elliptic cylinder with a strongly elongated cross section. Acoust. Phys. **61**(4), 383–387 (2015)
9. I.V. Andronov, R. Mittra, High-frequency diffraction by an elliptic cylinder: the far field. J. Electromagn. Waves Appl. **29**(10), 1317–1328 (2015)
10. I.V. Andronov, V.E. Petrov, Diffraction by an impedance strip at almost grazing incidence. IEEE Trans. Antennas Propag. **64**(8), 3565–3572 (2016)
11. R. Barakat, Diffraction of plane waves by an elliptic cylinder. J. Acoust. Soc. Am. **35**, 1990–1996 (1963)
12. J.J. Bowman, T.B.A. Senior, P.L.E. Uslenghi, *Electromagnetic and Acoustic Scattering by Simple Shapes* (North-Holland Publishing Co., Amsterdam, 1969)
13. H. Buchholz, *Die konfluente Hypergeometrische Functionen mit besonderer Berucksichtigung ihrer Anwendungen [The confluent hypergeometric functions with special reference to their applications]* (Springer-Verlag, Berlin, 1953)
14. R.F. Goodrich, N.D. Kazarinoff, Diffraction by thin elliptic cylinders. Michigan Math. J. **10**, 105–127 (1963)
15. H. Hönl, A. W. Maue, K. Westpfahl, *Theorie der Beugung* (Springer-Verlag, Berlin, 1961) *Theory of Diffraction* (Naval Intelligence Support Center, 1978)
16. V.I. Ivanov, Diffraction of short plane waves on a parabolic cylinder. USSR Comp. Math. Math. Phys. **2**, 255–271 (1963)
17. N.D. Kazarinoff, R.K. Ritt, On the theory of scalar diffraction and its application to the prolate spheroid. Ann. Phys. **6**, 277–299 (1959)
18. G.N. Kocherzhevski, Radiation of electric dipoles situated near a perfectly conducting elliptic cylinder. Zh. Tekhn. Fiz. **25**, 1140–1154 (1955)
19. I.V. Komarov, L.I. Ponomarev, *SYu* (Slavyanov, Spheroidal and Coulomb Spheroidal Functions (Science, Moscow, 1976)
20. B. Levy, Diffraction by an elliptic cylinder. J. Math. Mech. **9**, 147–166 (1960)
21. W.S. Lucke, Electric dipoles in the presence of elliptic and circular cylinders. J. Appl. Phys. **22**, 14–19 (1951)
22. F. Molinet, I. V. Andronov, Elliptic cylinder with a strongly elongated cross-section: high frequency techniques and function theoretic methods, in *Advances in Mathematical Methods for Electromagnetics* ed. by K. Kobayashi, P.D. Smith (the IET, London, UK, 2020), Chapter 6
23. J.W. Strutt, Lord Rayleigh, On the incidence of aerial and electric waves upon small obstacles in the form of ellipsoids or elliptic cylinders, and on the passage of electric waves through a circular aperture in a conducting screen. Philos. Mag. **44**, 28–52 (1897)
24. P.Y. Ufimtsev, *Fundamentals of the Physical Theory of Diffraction* (John Wiley & Sons, USA, 2007)

Chapter 3
Acoustic Wave Diffraction by Spheroid

3.1 Introduction

The simplest 3D model of an elongated body is the prolate spheroid. The case of scalar wave diffraction by a strongly elongated ellipsoid of revolution is considered in this chapter based on papers [2–7]. The solution u stands for the acoustic pressure and satisfies Helmholtz equation

$$\Delta u + k^2 u = 0. \tag{3.1}$$

The surface of the body is assumed to be well approximated by a prolate spheroid with the semiaxes b—the major and a—the minor. The spheroid is considered to be large and strongly elongated, that is

$$kb \gg 1, \qquad ka \gg 1, \qquad \text{and} \qquad (ka)^2 \sim kb.$$

We consider the following two cases of the boundary conditions on the surface S of the body. In the case of the absolutely soft surface, u satisfies the Dirichlet boundary condition

$$u|_S = 0.$$

In the case of the absolutely hard surface, u satisfies the Neumann boundary condition

$$\left.\frac{\partial u}{\partial n}\right|_S = 0,$$

where n is the outer normal to the surface S.

Two variants of the incident field u^{inc} are considered. In the first case, it is the plane wave incident at a small angle to the major axis of the spheroid. In the second case, the incident field is the spherical wave radiated by the point source. In the

I. Andronov, *Problems of High Frequency Diffraction by Elongated Bodies*, Springer Series in Optical Sciences 243, https://doi.org/10.1007/978-981-99-1276-6_3

latter case, we require that the position of the source with respect to the spheroid be such that the direction to the source and the major axis of the spheroid form a small angle. Besides, the source is assumed to be at a sufficiently large distance d from the surface, so that $kd \gg 1$. We also note that by adding small imaginary parts to the coordinates of the point source, one can obtain the problem of diffraction by a Gaussian beam. Such a problem is considered in [8].

First, we study the field in the boundary layer near the surface S. One can associate the field in the boundary layer with the waves that propagate along the geodesic lines on the surface. For the case of axial incidence, these geodesic lines coincide with the meridians connecting the poles at the tips of the spheroid. In this case, the system of geodesic lines is regular, but has focal points. In the case of skew incidence, the system of geodesic lines becomes complicated. However, the association of the field with the wave propagating along the geodesic line in the case of a strongly elongated body is not correct since the whole body lies in the Fock domain, and waves traveling along different trajectories do not have enough space to accumulate a phase difference. That means that the structure of the geodesic lines appears not important, and its complexity in the case of skew incidence does not cause any additional difficulties. We identify only two directions along the major axis of the spheroid: the forward direction on which the projection of the incident wave vector is positive and the opposite backward direction. The main contribution to the field in the boundary layer near the surface is evidently due to the wave process that propagates in the forward direction. This forward wave, when it reaches the rear end of the body, encircles it and launches waves that propagate backward. Finding the amplitudes of the backward waves requires matching of the forward and backward wave representation with the local asymptotic expansion which should be constructed in the vicinity of the rear tip. The latter faces the difficulty being in the fact that near the tips the curvature of the surface is large, so the radius of curvature occurs of the same order as the wavelength. This difficulty is overcome by using the approximation of the surface near the rear tip with an appropriate paraboloid. The interference of the forward and backward waves causes periodic oscillations of the field distribution along the surface.

The dependence on the transverse coordinate which accumulates the effect of skew incidence is represented by the angular Fourier harmonics, which have a similar structure to the forward and the backward fields. This enables us to present the derivations for all the harmonics at once. The need for a kind of Watson transform, which is applied to the series representation for the field in a usual canonical problem of diffraction, is eliminated by the assumption of the paraxial regime.

After getting the leading order approximation for the field in the boundary layer near the surface, by applying Green's formula we derive the far field representation. The forward wave produces the far field in the forward cone. The backward scattering effects are due to the two contributions. One is the specular reflection. Since the angle between the incident direction and the axis is assumed to be small, the specular point is close to the front tip of the spheroid, in the domain where the curvature of the surface is large. This is overcome by the same idea of using the approximation of the surface by a paraboloid, which is exploited in the procedure of finding the amplitudes

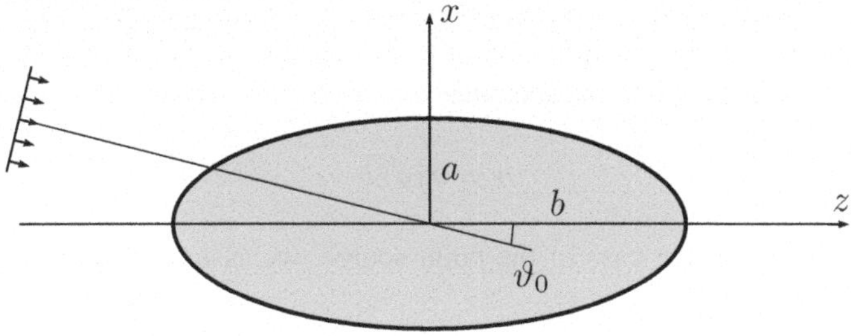

Fig. 3.1 Geometry of the problem

of the backward waves, but now the rear tip is considered. The other contribution to the backward scattered field is associated with the backward waves on the surface. The procedure of finding the radiation of the backward wave is similar to that for forward waves.

3.2 Problem Formulation and Assumptions on the Parameters

Figure 3.1 shows the geometry of the problem and illustrates the case of plane wave incidence. The Cartesian coordinate system is chosen such that the major axis of the spheroid coincides with the z axis, its origin is at the center of the spheroid and the plane of incidence is the Oxz plane. In the case of the point source field incidence, we assume that its y coordinate is equal to zero ($y_0 = 0$). It is convenient to use also the cylindrical coordinates (r, φ, z), such that

$$x = r\cos\varphi, \qquad y = r\sin\varphi$$

and spherical coordinates (R, ϑ, φ) such that

$$r = R\sin\vartheta, \qquad z = R\cos\vartheta.$$

We introduce the half of the inter-focal distance $p = \sqrt{b^2 - a^2}$ and use kp as the large asymptotic parameter. The rate of elongation is characterized by the quantity χ, introduced by the usual formula, which is assumed bounded

$$\chi = \frac{ka^2}{b} < const. \tag{3.2}$$

The above estimate is the requirement of the strong elongation of the spheroid; besides, we assume that the direction of plane wave incidence and the axis of the spheroid form a small angle ϑ_0, such that

$$\alpha = \vartheta_0 \sqrt{kb}$$

remains bounded. In the case of the point source, we assume that its coordinates (x_0, y_0, z_0) are such that

$$-kz_0 \gg kp, \qquad \xi = \frac{x_0^2}{|z_0 + b|} < const$$

and as was already mentioned $y_0 = 0$.

The total field in the problem is represented as the sum of the incident wave u^{inc} and the reflected wave u^{refl}, which in turn is the sum of the forward wave u^{forward} and the backward wave u^{back}. We specify the incident plane wave by the formula

$$u^{\text{inc}} = \exp\left(\mathrm{i}kz\cos\vartheta_0 + \mathrm{i}kx\sin\vartheta_0\right) \tag{3.3}$$

and the point source field as

$$u^{\text{inc}} = A\frac{\mathrm{e}^{\mathrm{i}kR}}{R}, \tag{3.4}$$

where $R = \sqrt{(z - z_0)^2 + y^2 + (x - x_0)^2}$.

Since u^{inc} is an exact solution of the Helmholtz equation, for u^{refl} we get the boundary-value problem

$$\Delta u^{\text{refl}} + k^2 u^{\text{refl}} = 0 \tag{3.5}$$

with the inhomogeneous boundary conditions

$$u^{\text{refl}}\Big|_{n=0} = -\,u^{\text{inc}}\Big|_{n=0} \tag{3.6}$$

(soft surface), or

$$\frac{\partial u^{\text{refl}}}{\partial n}\bigg|_{n=0} = -\,\frac{\partial u^{\text{inc}}}{\partial n}\bigg|_{n=0} \tag{3.7}$$

(hard surface). At infinity, u^{refl} satisfies the radiation conditions which can be specified in the form of the required asymptotic approximation

$$u^{\text{refl}}(\mathbf{R}) = \frac{\mathrm{e}^{\mathrm{i}kR}}{R}\Psi\left(\frac{\mathbf{R}}{R}\right) + O\left(R^{-2}\right), \qquad \|\mathbf{R}\| \equiv R \to +\infty. \tag{3.8}$$

3.3 Stretched Coordinates and Separation of Variables

Exploiting the symmetry of revolution of the surface, we represent the field in the form of Fourier series with respect to the angle φ in cylindrical coordinates (r, z, φ). Due to the symmetry of the incident wave, which is even with respect to y coordinate, we have

$$u = u_0(r, z) + 2\sum_{n=1}^{+\infty} u_n(r, z)\cos(n\varphi). \tag{3.9}$$

The incident plane wave is decomposed into series (3.9) with

$$u_n^{\text{inc}} = \mathrm{e}^n \exp(ikz\cos\vartheta_0) J_n\,(kr\sin\vartheta_0)\,,$$

where $J_n()$ are the Besselfunctions [1]. Every component u_n satisfies the corresponding equation

$$\frac{\partial^2 u_n}{\partial r^2} + \frac{1}{r}\frac{\partial u_n}{\partial r} + \frac{\partial^2 u_n}{\partial z^2} - \frac{n^2}{r^2}U_n + k^2 u_n = 0$$

and the boundary conditions (3.6) or (3.7).

We introduce spheroidal coordinates (η, ξ) by the formulas [20]

$$r = p\sqrt{1-\eta^2}\sqrt{\xi^2-1}, \quad z = p\eta\xi. \tag{3.10}$$

The surface of the spheroid is given then by the equation

$$\xi = \xi_0 \equiv \frac{b}{p}. \tag{3.11}$$

The Helmholtz equation for Fourier component u_n in spheroidal coordinates takes the form

$$\frac{\partial}{\partial\xi}\left(\left(\xi^2-1\right)\frac{\partial u_n}{\partial\xi}\right) + \frac{\partial}{\partial\eta}\left(\left(1-\eta^2\right)\frac{\partial u_n}{\partial\eta}\right)$$
$$-n^2\frac{\xi^2-\eta^2}{\left(\xi^2-1\right)\left(1-\eta^2\right)}u_n + \left(\xi^2-\eta^2\right)u_n = 0. \tag{3.12}$$

It is worth noting that the Helmholtz equation allows variable separation in spheroidal coordinates. So, the solutions to the problems of diffraction by soft or hard spheroids can be written in the form of infinite series containing spheroidal functions [11, 20]. However, such expressions appear ineffective in the case of high frequencies, and the resulting series should be subjected to a kind of Watson transform in order to obtain the asymptotic approximation. The use of spheroidal functions meets some additional difficulties as reported in [22] in the case of spheroids with large aspect ratio $b : a$. Therefore, we use a different approach and go further with the asymptotic simplifications of the problem. We note that in the case of a strongly elongated body, the following approximations hold

$$p \approx b - \frac{1}{2}\frac{\chi}{k}, \quad \xi_0 \approx 1 + \frac{1}{2}\frac{\chi}{kb}. \tag{3.13}$$

The last formula means that in the domain near the surface coordinate ξ takes the values, which are asymptotically close to one. Therefore, we introduce stretched coordinate τ

$$\tau = \frac{2kp}{\chi}(\xi - 1). \tag{3.14}$$

Similar to the usual parabolic equation method described in the first chapter, we extract quickly oscillating factor

$$u_n = \exp(\mathrm{i}kp\eta)U_n, \tag{3.15}$$

and search for the attenuation functions U_n in the form of the asymptotic series with respect to the inverse powers of large parameter kp, i.e.

$$U_n = \sum_{j=0}^{\infty} U_{nj}(\eta, \tau)(kp)^{-j}.$$

Inserting (3.15) into (3.12) and collecting the term of the similar orders in kp yields the recurrent system. In order to get a finite sequence of operators, we multiply beforehand the equation by $\xi^2 - 1$. Then it reduces to

$$\left(L_0 + \frac{\chi}{4kp}L_1 + \frac{\chi^2}{(4kp)^2}L_2\right)U_n = 0,$$

where

$$L_0 = \tau\frac{\partial^2}{\partial\tau^2} + \frac{\partial}{\partial\tau} + \frac{\mathrm{i}\chi}{2}\left(1-\eta^2\right)\frac{\partial}{\partial\eta} + \left(\frac{\chi^2}{4}\tau - \frac{\mathrm{i}\chi\eta}{2} - \frac{n^2}{4\tau}\right),$$

$$L_1 = 2\tau^2\frac{\partial^2}{\partial\tau^2} + 3\tau\frac{\partial}{\partial\tau} + \left(1-\eta^2\right)\frac{\partial^2}{\partial\eta^2} + \left[\frac{\mathrm{i}\chi\tau}{2}\left(1-\eta^2\right) - 2\eta\right]\frac{\partial}{\partial\eta} + \left[\frac{\chi^2}{2}\tau - \frac{\mathrm{i}\chi}{2}\eta\tau - \frac{n^2}{2\left(1-\eta^2\right)}\right],$$

$$L_2 = \tau^3 \frac{\partial^2}{\partial \tau^2} + 2\tau^2 \frac{\partial}{\partial \tau} + \left(1 - \eta^2\right) \tau \frac{\partial^2}{\partial \eta^2} - 2\eta\tau \frac{\partial}{\partial \eta} + \tau \left[\frac{\chi^2}{4}\tau^2 - \frac{n^2}{1 - \eta^2} \right].$$

Thus, functions U_{nj} satisfy the recurrent system of equations

$$L_0 U_{n0} = 0, \qquad L_0 U_{n1} = -\frac{\chi}{4} L_1 U_{n0}, \qquad \ldots$$

which allows, in principle, to determine step by step the terms of the asymptotic series up to any desired order. We restrict our derivations to finding only the leading order term U_{n0} and suppress below the index $j = 0$.

3.4 The Forward Wave

3.4.1 *Integral Representation*

At the leading order, the attenuation functions U_n satisfy the parabolic equations (we substitute here operator L_0 explicitly):

$$\tau \frac{\partial^2 U_n}{\partial \tau^2} + \frac{\partial u_n}{\partial \tau} + \frac{\mathrm{i}\chi}{2} \left(1 - \eta^2\right) \frac{\partial u_n}{\partial \eta} + \left(\frac{\chi^2}{4}\tau - \frac{n^2}{4\tau} - \frac{\mathrm{i}\chi}{2}\eta \right) U_n = 0. \tag{3.16}$$

The parabolic equations (3.16) are very similar to those in Chap. 2; see (2.7).

The general solution of (3.16) can be represented in the form of the integral by parameter t of variable separation. Substituting instead of U_n the product $\tau^{-1/2} F(-\mathrm{i}\chi\tau) G(\eta)$, we get from (3.16) Whittaker equation

$$F''(s) + \left[-\frac{1}{4} + \frac{\mathrm{i}t}{s} + \frac{1 - n^2}{4s^2} \right] F(s) = 0$$

for function F and the first-order differential equation

$$G'(\eta) = \frac{\eta - 2\mathrm{i}t}{1 - \eta^2} G(\eta)$$

for function G. Thus, F is a combination of Whittaker functions $M_{\mathrm{i}t, n/2}$ and $W_{\mathrm{i}t, n/2}$ and

$$G(\eta) = \frac{1}{\sqrt{1 - \eta^2}} \left(\frac{1 - \eta}{1 + \eta} \right)^{\mathrm{i}t}.$$

We had a very similar function G in Chap. 2 in (2.9). So, by using similar argumentation we can write the integral representation for functions U_n in the form

$$U_n = \frac{1}{\sqrt{\chi\tau}\sqrt{1-\eta^2}} \int\limits_{-\infty}^{+\infty} \left(\frac{1-\eta}{1+\eta}\right)^{it} A_n(t) \times \\ \times \left\{M_{it,n/2}(-i\chi\tau) + R_n(t) W_{it,n/2}(-i\chi\tau)\right\} dt. \quad (3.17)$$

The coefficients A_n and R_n at this step of derivations are arbitrary. Similar to Chap. 2, we note that

$$U_n^{\text{inc}} = \frac{1}{\sqrt{\chi\tau}\sqrt{1-\eta^2}} \int\limits_{-\infty}^{+\infty} \left(\frac{1-\eta}{1+\eta}\right)^{it} A_n(t) M_{it,n/2}(-i\chi\tau) dt \quad (3.18)$$

and

$$U_n^{\text{refl}} = \frac{1}{\sqrt{\chi\tau}\sqrt{1-\eta^2}} \int\limits_{-\infty}^{+\infty} \left(\frac{1-\eta}{1+\eta}\right)^{it} A_n(t) R_n(t) W_{it,n/2}(-i\chi\tau) dt.$$

Thus, coefficients A_n depend only on the incident field, and an appropriate choice of coefficients R_n allows the boundary conditions to be satisfied. Namely, in the case of the soft surface we take

$$R_n(t) = -\frac{M_{it,n/2}(-i\chi)}{W_{it,n/2}(-i\chi)}, \quad (3.19)$$

and in the case of the hard surface

$$R_n(t) = -\frac{M_{it,n/2}(-i\chi) + 2i\chi \dot{M}_{it,n/2}(-i\chi)}{W_{it,n/2}(-i\chi) + 2i\chi \dot{W}_{it,n/2}(-i\chi)}, \quad (3.20)$$

where the dot denotes the derivative of the function.

3.4.2 *The Incident Plane Wave Representation*

In this section, we consider the case of the incident plane wave (3.3). In order to find coefficients A_n in representation (3.18), we consider the incident wave in the boundary layer near the surface and rewrite it using the boundary-layer coordinates. Discarding smaller order terms, we get

$$kz\cos\vartheta_0 \approx kp\eta + \frac{\chi\tau - \alpha^2}{2}\eta, \quad (3.21)$$

$$kr\sin\vartheta_0 \approx \alpha\sqrt{\chi\tau}\sqrt{1-\eta^2}. \tag{3.22}$$

Equating Fourier components u_n^{inc} of the incident wave with the representation (3.17), we get the following equation:

$$\mathrm{i}^n \exp\left(\frac{\mathrm{i}}{2}\left(\chi\tau-\alpha^2\right)\eta\right) J_n\left(\sqrt{\chi\tau}\alpha\sqrt{1-\eta^2}\right) = \frac{1}{\sqrt{\chi\tau}\sqrt{1-\eta^2}}\int\limits_{-\infty}^{+\infty}\left(\frac{1-\eta}{1+\eta}\right)^{\mathrm{i}t} A_n(t) M_{\mathrm{i}t,n/2}(-\mathrm{i}\chi\tau)\mathrm{d}t. \tag{3.23}$$

The coefficients A_n do not depend on variable τ, but this variable is presented in formula (3.23) only as the product $\chi\tau$, therefore one can conclude that A_n do not depend on χ either. Further, α^2 and $-\chi\tau$ are presented on the left-hand side of formula (3.23) in a symmetric way, therefore after symmetrizing the right-hand side by setting

$$A_n = \frac{M_{\mathrm{i}t,n/2}(\mathrm{i}\alpha^2)}{\alpha} a_n, \tag{3.24}$$

we get new unknowns a_n that do not depend on α as well. So, coefficients a_n depend only on n and t. These coefficients can be found from (3.23), where one can choose convenient values of parameters $\chi\tau$ and α. Letting $\chi\tau \to 0$ and $\alpha \to 0$ and noting that [1]

$$J_n(x) \sim \frac{1}{n!}\left(\frac{x}{2}\right)^n, \quad x \to 0 \tag{3.25}$$

and [16, formula 9.226]

$$M_{\mu,m}(x) \sim x^{1/2+m}, \quad x \to 0, \tag{3.26}$$

we get the much more simple equation

$$\mathrm{i}^n \frac{\left(1-\eta^2\right)^{(n+1)/2}}{2^n n!} = \int\limits_{-\infty}^{+\infty}\left(\frac{1-\eta}{1+\eta}\right)^{\mathrm{i}t} a_n(t)\mathrm{d}t. \tag{3.27}$$

This equation is equivalent to the former (3.23). That is, one can find the solution of (3.27) and define A_n according to formula (3.24). Such A_n satisfy (3.23) identically for any positive values of χ, τ and α.

The solution of (3.27) can be obtained by using the transform expressed by formulas (A.32) and (A.33). We get

$$a_n = \frac{\mathrm{i}^n}{\pi 2^n n!}\int\limits_{-1}^{1}\left(\frac{1+\eta}{1-\eta}\right)^{\mathrm{i}t}\left(1-\eta^2\right)^{(n-1)/2}\mathrm{d}\eta.$$

To rewrite this integral in terms of special functions, one changes the integration variable to

$$s = \frac{1+\eta}{1-\eta}.$$

Since

$$\eta = 1 - \frac{2}{s+1}, \qquad 1-\eta^2 = \frac{4s}{(s+1)^2}, \qquad \mathrm{d}\eta = \frac{2}{(s+1)^2},$$

we get

$$a_n = \frac{\mathrm{i}^n}{\pi n!} \int\limits_0^{+\infty} \frac{s^{\mathrm{i}t+\frac{n-1}{2}}}{(s+1)^{n+1}} \mathrm{d}s.$$

Comparing this integral with the representation for the beta-function, we finally get

$$a_n = \frac{\mathrm{i}^n}{\pi (n!)^2} \Gamma\left(\frac{n+1}{2} + \mathrm{i}t\right) \Gamma\left(\frac{n+1}{2} - \mathrm{i}t\right).$$

Thus,

$$A_n = \frac{\mathrm{i}^n}{\pi (n!)^2} \Gamma\left(\frac{n+1}{2} + \mathrm{i}t\right) \Gamma\left(\frac{n+1}{2} - \mathrm{i}t\right) \frac{M_{\mathrm{i}t,n/2}(\mathrm{i}\alpha^2)}{\alpha}. \tag{3.28}$$

In the case of axial incidence, $\alpha = 0$ and one needs to exclude the ambiguity. Using the approximation (3.26), we get

$$A_0 = \frac{\mathrm{e}^{\mathrm{i}\pi/4}}{\cosh(\pi t)}, \qquad A_n = 0, \quad n = 1, 2, 3, \ldots \tag{3.29}$$

3.4.3 The Incident Spherical Wave Representation

In the case of point source field incidence, we represent the incident field (3.4) in the form of Fourier series,

$$u^{\mathrm{inc}} = \frac{1}{2} g_0 + \sum_{n=1}^{\infty} g_n \cos(n\varphi),$$

where

$$g_n = \frac{A}{\pi} \int\limits_0^{2\pi} u^{\mathrm{inc}}(r, \varphi, z; r_0, z_0) \cos(n\varphi) \mathrm{d}\varphi, \quad n = 0, 1, 2, \ldots \tag{3.30}$$

Reminding that point $\mathbf{R}_0$ is assumed to be located in a narrow cone near the axis of the spheroid and considering also $r \ll |z - z_0|$, we come to the approximation

$$R \approx (z - z_0)\left[1 + \frac{1}{2}\frac{r^2 + r_0^2 - 2rr_0\cos(\varphi)}{(z - z_0)^2}\right].$$

Then the integrals in (3.30) can be reduced to Besselfunctions, and we get

$$g_n \approx 2A\mathrm{i}^n \frac{\mathrm{e}^{\mathrm{i}k(z-z_0)}}{z - z_0}\exp\left(\frac{\mathrm{i}k}{2}\frac{r^2 + r_0^2}{z - z_0}\right) J_n\left(-\mathrm{i}k\frac{rr_0}{z - z_0}\right). \tag{3.31}$$

Analogously to (3.15), we extract the multiplier by introducing the attenuation functions G_n as $g_n = \mathrm{e}^{\mathrm{i}kp\eta}G_n$ and rewrite the right-hand side of formula (3.31) in terms of the boundary-layer coordinates (η, τ):

$$G_n = \frac{2A\mathrm{i}^n\mathrm{e}^{-\mathrm{i}kz_0}}{b(\eta + \zeta_0)}\exp\left(\frac{\mathrm{i}\chi}{2}\tau\eta + \frac{\mathrm{i}\chi}{2}\frac{(1 - \eta^2)\tau + \gamma_0^2}{\eta + \zeta_0}\right) \times$$
$$\times J_n\left(\chi\frac{\sqrt{1 - \eta^2}\sqrt{\tau}\gamma_0}{\eta + \zeta_0}\right). \tag{3.32}$$

Here, the dimensionless coordinates of the source are introduced as

$$\gamma_0 = \frac{r_0}{a}, \qquad \zeta_0 = -\frac{z_0}{b}.$$

One can check that functions G_n given by (3.32) satisfy the parabolic equations $L_0G_n = 0$ and therefore can be rewritten in the form of representation (3.18), which enables us to write the equations

$$\int_{-\infty}^{+\infty}\left(\frac{1 - \eta}{1 + \eta}\right)^{\mathrm{i}t} A_n(t)M_{\mathrm{i}t,n/2}(-\mathrm{i}\chi\tau)\mathrm{d}t = A\sqrt{\chi\tau}\sqrt{1 - \eta^2}G_n(\eta, \tau) \tag{3.33}$$

for functions A_n.

Noting the transform (A.32) in (3.33) and applying the inverse transform (A.33), we can rewrite functions A_n in integral form. However, in order to simplify the resulting expressions and in order to rewrite A_n in terms of special functions, we first perform similar transformations as in the case of plane wave incidence. Setting τ arbitrarily small, in view of the asymptotic approximations (3.25) and (3.26), we replace $M_{\mathrm{i}t,n/2}(-\mathrm{i}\chi\tau)$ with $(-\mathrm{i}\chi\tau)^{1/2+n/2}$ and

$$J_n\left(\chi\frac{\sqrt{1 - \eta^2}\sqrt{\tau}\gamma_0}{\eta + \zeta_0}\right) \quad \text{with} \quad \frac{1}{2^n n!}\left(\chi\frac{\sqrt{1 - \eta^2}\sqrt{\tau}\gamma_0}{\eta + \zeta_0}\right)^n.$$

This simplifies (3.33) to

$$\int_{-\infty}^{+\infty} \left(\frac{1-\eta}{1+\eta}\right)^{it} A_n(t)\mathrm{d}t = C_n \left[\frac{\sqrt{1-\eta^2}}{\eta+\zeta_0}\right]^{n+1} \exp\left(\frac{\mathrm{i}\chi}{2}\frac{\gamma_0^2}{\eta+\zeta_0}\right),$$

where

$$C_n = A(-1)^n \frac{\mathrm{e}^{\mathrm{i}\pi(1-n)/4}}{2^{n-1}n!}\frac{\mathrm{e}^{-\mathrm{i}kz_0}}{b}\left[\sqrt{\chi}\gamma_0\right]^n$$

does not depend on η. Now we use the inverse transform (A.33)

$$A_n = \frac{C_n}{\pi}\int_{-1}^{1}\left(\frac{1+\eta}{1-\eta}\right)^{it}\frac{(1-\eta^2)^{(n-1)/2}}{(\eta+\zeta_0)^{n+1}}\exp\left(\frac{\mathrm{i}}{2}\frac{\chi\gamma_0^2}{\eta+\zeta_0}\right)\mathrm{d}\eta$$

and after introducing a new variable $s_1 = (1+\eta)/(1-\eta)$ obtain the following representation

$$A_n = \frac{C_n}{2\pi}\int_{0}^{+\infty}\left[\frac{1}{\zeta_0-1}\frac{2\sqrt{s_1}}{s_1\theta+1}\right]^{n+1}\exp\left(\frac{\mathrm{i}}{2}\frac{\chi\gamma_0^2}{\zeta_0-1}\frac{s_1+1}{s_1\theta+1}\right)s_1^{\mathrm{i}t-1}\mathrm{d}s_1,$$

where $\theta = (\zeta_0+1)/(\zeta_0-1)$. Further, by introducing the integration variable $s_2 = 2/(s_1\theta+1)-1$, the above integral can be rewritten as

$$A_n = \frac{C_n}{\pi}(\zeta_0+1)^{-\frac{n+1}{2}-\mathrm{i}t}(\zeta_0-1)^{-\frac{n+1}{2}+\mathrm{i}t}\exp\left(\frac{\mathrm{i}}{2}\Upsilon_0\zeta_0\right)\times$$
$$\times\int_{-1}^{1}(s_2+1)^{\frac{n-1}{2}-\mathrm{i}t}(1-s_2)^{\frac{n-1}{2}+\mathrm{i}t}\exp\left(\frac{\mathrm{i}\Upsilon_0 s_2}{2}\right)\mathrm{d}\xi, \qquad (3.34)$$

where

$$\Upsilon_0 = \frac{\chi\gamma_0^2}{\zeta_0^2-1}.$$

Identifying the integral in (3.34) with the integral representation (A.9) for the Whittaker function M, we finally get

$$A_n = \frac{2A\,\mathrm{i}^n}{\pi(n!)^2}\frac{\mathrm{e}^{-\mathrm{i}kz_0}}{b\sqrt{\chi}\gamma_0}\exp\left(\frac{\mathrm{i}\Upsilon\zeta_0}{2}\right)\left(\frac{\zeta_0-1}{\zeta_0+1}\right)^{it}\times$$
$$\times\Gamma\left(\frac{n+1}{2}+\mathrm{i}t\right)\Gamma\left(\frac{n+1}{2}-\mathrm{i}t\right)M_{\mathrm{i}t,n/2}(\mathrm{i}\Upsilon_0). \qquad (3.35)$$

We remark that expression (3.35) is the exact solution of the integral equation (3.33), that is, substituting from (3.35) to (3.33) makes the identity, which is satisfied for any values of $\tau \geq 0$.

In the case of $\gamma_0 = 0$, excluding the ambiguity in (3.35) we get

$$A_0 = \mathrm{e}^{\mathrm{i}\pi/4} \frac{\mathrm{e}^{-\mathrm{i}kz_0}}{b\sqrt{\eta_0^2 - 1}} \left(\frac{\zeta_0 - 1}{\zeta_0 + 1} \right)^{\mathrm{i}t} \frac{1}{\cosh(\pi t)}$$

and $A_n = 0$ for $n = 1, 2, \ldots$.

3.4.4 The Currents

Substituting expression (3.28) to (3.17) gives the leading order approximation for the field in the boundary layer near the surface in the case of plane wave incidence. In particular, one can find the field on the surface. For that, we set $\tau = 1$ and use the expression for the Wronskian of Whittaker functions (A.21)

$$\dot{M}_{\mathrm{i}t,n/2} W_{\mathrm{i}t,n/2} - \dot{W}_{\mathrm{i}t,n/2} M_{\mathrm{i}t,n/2} = \frac{n!}{\Gamma\left(\frac{n+1}{2} - \mathrm{i}t\right)}.$$

In the case of the soft surface, we find

$$\left.\frac{\partial U_n}{\partial \tau}\right|_{\tau=1} = \frac{\mathrm{i}^n}{\pi n!} \frac{1}{\sqrt{\chi}\alpha\sqrt{1-\eta^2}} \int\limits_{-\infty}^{+\infty} \left(\frac{1-\eta}{1+\eta} \right)^{\mathrm{i}t} \Gamma\left(\frac{1+n}{2} + \mathrm{i}t \right) \times \\ \times \frac{M_{\mathrm{i}t,n/2}(\mathrm{i}\alpha^2)}{W_{\mathrm{i}t,n/2}(-\mathrm{i}\chi)} \mathrm{d}t. \qquad (3.36)$$

On the hard surface, we get

$$U_n|_{\tau=1} = -\frac{\mathrm{i}^n}{\pi n!} \frac{1}{\sqrt{\chi}\alpha\sqrt{1-\eta^2}} \int\limits_{-\infty}^{+\infty} \left(\frac{1-\eta}{1+\eta} \right)^{\mathrm{i}t} \Gamma\left(\frac{1+n}{2} + \mathrm{i}t \right) \times \\ \times \frac{M_{\mathrm{i}t,n/2}(\mathrm{i}\beta_0^2)}{\dot{W}_{\mathrm{i}t,n/2}(-\mathrm{i}\chi) - \frac{\mathrm{i}}{2\chi} W_{\mathrm{i}t,n/2}(-\mathrm{i}\chi)} \mathrm{d}t. \qquad (3.37)$$

The field of normal fluid-particle velocities on a soft surface can be found by summing up Fourier components (3.36) and multiplying the result by the quick factor $\exp(\mathrm{i}kp\eta)$. Similarly, the surface potential on a hard surface is given by the sum of harmonics (3.37) multiplied by the quick factor. The questions of convergence of these series and the smoothness of the solution are studied in [9]. For $\alpha = 0$ in view of formula (3.29), the only nonzero Fourier component is the one with $n = 0$ and in formulas (3.36) and (3.37) expression $M_{\mathrm{i}t,0}(\mathrm{i}\alpha^2)/\alpha$ should be replaced with $\mathrm{e}^{\mathrm{i}\pi/4}$, which follows from (A.22).

The analysis of formulas (3.36) and (3.37) shows that the leading order terms of the asymptotic expansions have singularities at the shadowed end-point of the spheroid, at $\eta = 1$. Though we do not present the next order terms of the asymptotic series, one checks that the order of singularities increases with j. Therefore, asymptotic formulas (3.36) and (3.37) fail when $\eta \to 1$. In a similar way, these formulas are not valid in the vicinity of the illuminated end-point, when $\eta \to -1$. We shall examine the field in the vicinity of the shadowed end-point in the next section, and the illuminated end-point vicinity is examined in Sect. 3.7.3.

3.5 Vicinities of the End-points and the Backward Diffracted Wave

The field in the vicinities of the spheroid end-points is not described by the asymptotic expansions derived in the previous sections because the field in these domains should not contain the oscillating factor $\mathrm{e}^{\mathrm{i}kp\eta}$ which is extracted in these asymptotic formulas. The asymptotic analysis of the diffracted fields in these domains requires special considerations. The difficulty comes from the observation that the radius of curvature of the surface ρ is not large here, so $k\rho = O(1)$. Indeed, at the end-points $\rho = a^2/b$, which means that $k\rho = \chi$. Therefore, the usual high-frequency asymptotic methods are not applicable. However, there is a way out. It is based on the approximation of the surface of the spheroid at these domains with the surface of an appropriate paraboloid.

We consider in this section only the vicinity of the shadowed end-point. There we introduce paraboloidal coordinates (v, w) such that

$$z = p + \frac{v - w}{2k}, \qquad r = \frac{\sqrt{vw}}{k}. \tag{3.38}$$

Then the surface of the spheroid is approximated by the surface $v = \chi$.

Equation (3.1) in paraboloidal coordinatesreduces to

$$4w\frac{\partial^2 U}{\partial w^2} + 4\frac{\partial U}{\partial w} + \left(w - \frac{n^2}{w}\right)U + +4v\frac{\partial^2 U}{\partial v^2} + 4\frac{\partial U}{\partial w} + \left(v - \frac{n^2}{v}\right)U = 0.$$

Its solution can be written in the form of the integral [14, 18]

$$U_n = \frac{1}{\sqrt{vw}} \int B_n(t) M_{\mathrm{i}t,n/2}(\mathrm{i}w) W_{\mathrm{i}t,n/2}(-\mathrm{i}v)\mathrm{d}t. \tag{3.39}$$

Here, the choice of Whittaker function $M_{it,n/2}(iw)$ is determined by the requirement that the field U_n should be regular at $w = 0$. The choice of Whittaker function $W_{it,n/2}(-iv)$ is determined by the radiation condition for $v \to +\infty$. The path of integration and the coefficients B_n are as far as arbitrary. We define these coefficients by requiring that expressions (3.39) match the asymptotic expansions of Sect. 3.4 far from the end-point, that is for $w \to +\infty$.

Under the assumption of (3.2), we find the relations between the stretched spheroidal coordinates (η, τ) and paraboloidal coordinates (v, w). Discarding smaller order terms, we can write these relations as

$$v = \chi\tau, \qquad w = 2kp(1-\eta). \tag{3.40}$$

Then the forward wave asymptotic approximation reduces in the vicinity of the shadowed end-point to

$$U_n^{\text{forward}} = \frac{\sqrt{kp}\,e^{ikp-iw/2}}{\sqrt{vw\alpha}} \times \\ \times \int\limits_{-\infty}^{+\infty} w^{it}(4kp)^{-it} a_n(t) M_{it,n/2}(i\alpha^2) R_n(t) W_{it,n/2}(Iv)\,dt.$$

In formula (3.39), we consider $w \to +\infty$ and replace Whittaker function $M_{it,n/2}(iw)$ with its asymptotics (A.24)

$$M_{it,n/2}(iw) \sim \exp\left(-\frac{iw}{2} \pm i\pi\left(\frac{n+1}{2} - it\right)\right)\frac{n!}{\Gamma\left(\frac{n+1}{2}+it\right)}(iw)^{it} \\ + \exp\left(\frac{iw}{2}\right)\frac{n!}{\Gamma\left(\frac{n+1}{2}-it\right)}(iw)^{-it}. \tag{3.41}$$

Thus, for $w \to +\infty$ we get

$$U_n \sim \frac{n!}{\sqrt{vw}}\int B_n(t)\left\{e^{-iw/2} i^{n+1} e^{\pi t}\frac{(iw)^{it}}{\Gamma\left(\frac{n+1}{2}+it\right)}\right. \\ \left. + e^{iw/2}\frac{(iw)^{-it}}{\Gamma\left(\frac{n+1}{2}-it\right)}\right\} W_{it,n/2}(-iv)\,dt. \tag{3.42}$$

It is easy to see that the forward wave matches the first term in the braces in the above formula. This allows the coefficients B_n to be found. We have

$$B_n = \sqrt{kb}(4kb)^{-it} e^{ikp}\frac{M_{it,n/2}(i\alpha^2)}{\alpha}\frac{\Gamma\left(\frac{n+1}{2}+it\right)}{n!\,i^{n+1}} a_n(t). \tag{3.43}$$

The second term in (3.42) evidently forms the backward field which is of primary interest. In the middle part of the spheroid, this backward field can be represented in the form similar to (3.17) with η replaced with $-\eta$ which is the consequence of the opposite direction of propagation. Also there is no backward incident wave, that is,

$$U_n^{\text{back}} = \frac{\mathrm{e}^{-\mathrm{i}kp\eta}}{\sqrt{\chi\tau}\sqrt{1-\eta^2}} \int_{-\infty}^{+\infty} \left(\frac{1+\eta}{1-\eta}\right)^{it} C_n(t) R_n(t) W_{\mathrm{i}t,n/2}(-\mathrm{i}\chi\tau)\mathrm{d}t. \tag{3.44}$$

By matching this formula for $\eta \to 1$ with the part of expression (3.42) with the second term in the braces, we find the coefficients

$$C_n = A_n r_n(t),$$

$$r_n(t) = \mathrm{e}^{2\mathrm{i}kp}(4kb)^{-2\mathrm{i}t}(-\mathrm{i})^{n+1} \frac{\Gamma\left(\frac{n+1}{2}+\mathrm{i}t\right)}{\Gamma\left(\frac{n+1}{2}-\mathrm{i}t\right)}. \tag{3.45}$$

Functions r_n play the role of partial reflection coefficients of the forward diffracted waves from the rear ending of the spheroid. These coefficients quickly oscillate due to the multiplier $(kb)^{-2\mathrm{i}t}$. Therefore, the path of integration in (3.44) can be shifted to the upper half-plane of complex t, which reduces the integral to the sum of residues in the poles defined by the equations

$$W_{\mathrm{i}t,n/2}(-\mathrm{i}\chi) = 0 \tag{3.46}$$

in the case of soft surface, and

$$\dot{W}_{\mathrm{i}t,n/2}(-\mathrm{i}\chi) - \frac{\mathrm{i}}{2\chi} W_{\mathrm{i}t,n/2}(-\mathrm{i}\chi) = 0 \tag{3.47}$$

in the case of hard surface. The solutions of dispersion equations (3.46) and (3.47) as functions of χ are studied in Appendix A.3.

3.6 Numerical Results and Validations for the Currents

Combining the above-derived expressions, we get the approximation for the field in the boundary layer near the surface

$$u = \frac{4\sqrt{\chi}}{\pi\sqrt{1-\eta^2}} \sum_{n=0}^{+\infty} \frac{\mathrm{i}^n \cos(n\varphi)}{n!(1+\delta_n^0)} \int_{-\infty}^{+\infty} \left(\frac{1-\eta}{1+\eta}\right)^{\mathrm{i}t} \Gamma\left(\frac{n+1}{2}+\mathrm{i}t\right) \times$$
$$\times \frac{M_{\mathrm{i}t,n/2}(\mathrm{i}\alpha^2)}{\alpha} \frac{\mathrm{e}^{\mathrm{i}kp\eta} + r_n(t)\mathrm{e}^{-\mathrm{i}kp\eta}}{\mathrm{i}W_{\mathrm{i}t,n/2}(-\mathrm{i}\chi) - 2\chi \dot{W}_{\mathrm{i}t,n/2}(-\mathrm{i}\chi)} \mathrm{d}t. \quad (3.48)$$

Computations according to this formula show that the number of the Fourier harmonics, which give significant contributions to the field, is not large. For axial incidence, one has only the zeroth-order harmonics. For the case of skew incidence, it is usually sufficient to take no more than 10 harmonics. The integrated expressions rapidly decrease at $t \to \pm\infty$. The size of the interval which contributes to the integral depends on the values of parameters χ and α. For small χ and α, one can restrict the domain of integration to the interval $[-3, 3]$. When χ increases, the lower bound should be shifted to the left. For example, for $\chi = 5$ the interval becomes $[-6, 2]$ and for $\chi = 15$— $[-9, 2]$. The increase of α leads to the necessity to shift rightward the upper bound of the interval. For example in the case of $\alpha = 3$, the right-hand bound can be accepted at $t = 6$. Whittaker functions W and M can be computed with the help of the COULCC routine [21], which computes the Coulomb wave functions F_L and H_L^+.

Figure 3.2 illustrates the pressure distribution on the surface of hard spheroids. The wave size kb is fixed to the value of 20 while the rate of elongation is taken different, so that χ be equal to 0.5, 1, 2, 5 and 10. The pressure crucially depends on the rate of elongation. When χ is large, the pressure quickly decreases along the surface, and in the limiting case of $\chi \to \infty$ its distribution transforms to the Fock asymptotic approximation (see Chap. 1). When the elongation parameter is moderate and small, that is the body is sufficiently much elongated, the amplitude of the pressure at the illuminated side of the body is smaller, but simultaneously the rate of attenuation along the surface decreases. This results in higher values of $|u|$ at the rear (shadowed) side of the body. The graphics in Fig. 3.2 show that the contribution of the backward Wave is noticeable only near the rear end of the body, and it rapidly decreases when the observation point moves to the illuminated part of the surface.

The asymptotic approximation for the field on a hard spheroid can be compared with numerical simulations. The boundary integral methods in their traditional versions are restricted to not too high frequencies. Below, we present the results taken from [12] where the multilevel fast-multipole algorithm on the multilevel nonuniform grid [13] is validated by comparing the currents computed by this algorithm with the asymptotic approximation (3.48). The tests have been performed for the two hard spheroids. Both having the length of 10 m, but different aspect ratios 1:5 (spheroid A) and 1:10 (spheroid B). The values of the acoustic pressures on the surface excited by a plane wave at different frequencies and angles of incidence have been compared. We denote by u_a the asymptotic approximation expressed by (3.48), and by u_f the numerical results for the field.

The fast solver computed the fields at the centroids of the mesh, which have been chosen by the numerical algorithm. The asymptotic approximations can be

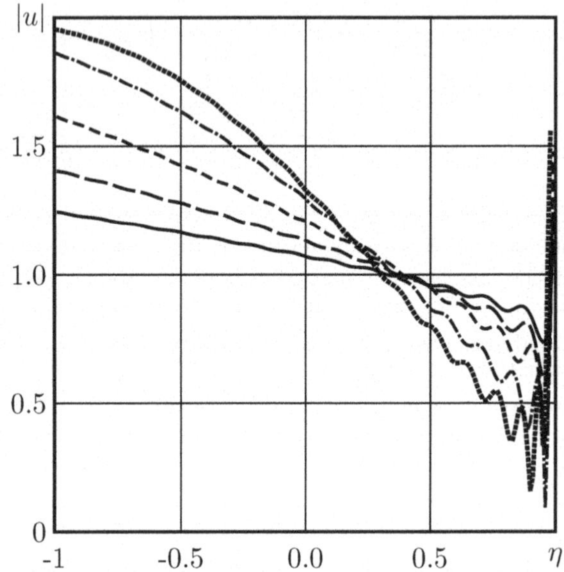

Fig. 3.2 Forward and backward waves on the surface of hard spheroids with $kb = 20$ and different elongations. Solid line $\chi = 0.5$, long dashed $\chi = 1$, short dashed $\chi = 2$, dash-dotted $\chi = 5$ and dotted $\chi = 10$

computed exactly at the same points on the surface and then the differences have been calculated. We compare the fields only in the middle part of spheroids, where the asymptotic approximations developed in this chapter are valid. We present the results for both the amplitudes of the fields, that is $\delta_1 = |u_a| - |u_f|$, and for the complex values $\delta_2 = |u_a - u_f|$. So, the difference δ_2 includes the highly oscillatory factor and therefore is more sensitive to the parameter kp. Table 3.1

presents the average errors. In order to obtain correct asymptotic approximation for the field on Spheroid A at the highest numerically accessible frequency for the incidence at 10°, the number of angular harmonics taken into account have been increased to 20. For all the other cases, it was set equal to 10. The difference between the surface fields calculated using the fast solver and the asymptotic approximation (3.48) is generally within the error of numerical computations, estimated as 1–2%.

3.7 The Far Field

3.7.1 Kirchhoff Integral

The far field is expressed by Kirchhoff integral [17], into which we substitute the derived above asymptotic approximation for the field in the boundary layer.

Table 3.1 The average error values at different frequencies

Spheroid A				$\vartheta = 0°$		$\vartheta = 10°$	
n	kp	χ	Centroids	$\|\delta_1\|$	$\|\delta_2\|$	$\|\delta_1\|$	$\|\delta_2\|$
1	30.781	1.283	11232	0.0072	0.0192	0.0077	0.0207
2	43.531	1.814	21624	0.0078	0.0217	0.0082	0.0235
3	61.562	2.565	42076	0.0082	0.0231	0.0083	0.0255
4	87.062	3.628	85816	0.0081	0.0268	0.0082	0.0299
5	123.125	5.130	169344	0.0085	0.0283	0.0087	0.0329
6	174.125	7.255	328852	0.0075	0.0433	0.0084	0.0361
7	246.250	10.260	644252	0.0087	0.0419	0.0085	0.0420
8	348.249	14.510	1261324	0.0088	0.0720	0.0079	0.0327
Spheroid B				$\vartheta = 0°$		$\vartheta = 10°$	
n	kp	χ	Centroids	$\|\delta_1\|$	$\|\delta_2\|$	$\|\delta_1\|$	$\|\delta_2\|$
1	31.258	0.316	10486	0.0019	0.0049	0.0028	0.0064
2	44.206	0.447	10486	0.0036	0.0074	0.0036	0.0076
3	62.517	0.631	44180	0.0017	0.0040	0.0021	0.0063
4	88.412	0.893	44180	0.0032	0.0060	0.0029	0.0063
5	125.034	1.263	88124	0.0038	0.0074	0.0038	0.0123
6	176.825	1.786	174530	0.0039	0.0077	0.0037	0.0085
7	250.068	2.526	355243	0.0041	0.0088	0.0051	0.0204
8	353.649	3.572	714224	0.0039	0.0090	0.0044	0.0106
9	500.135	5.052	1476190	0.0048	0.0111	0.0059	0.0362

$$u^{\text{scat}}(\mathbf{R}_0) = \frac{1}{4\pi} \int_S \left(G(\mathbf{R}(s), \mathbf{R}_0) \frac{\partial u^{\text{refl}}(\mathbf{R}(s))}{\partial n} - \frac{\partial G(\mathbf{R}(s), \mathbf{R}_0)}{\partial n} u^{\text{refl}}(\mathbf{R}(s)) \right) ds. \tag{3.49}$$

Here, $G(\mathbf{R}, \mathbf{R}_0)$ is Green's function in free space

$$G(\mathbf{R}, \mathbf{R}_0) = \frac{e^{ik|\mathbf{R}-\mathbf{R}_0|}}{|\mathbf{R} - \mathbf{R}_0|}, \tag{3.50}$$

$u^{\text{refl}}(\mathbf{R})$ is the reflected (secondary) field, S is the surface of the spheroid and n is the normal to S. Instead of the secondary field u^{refl}, one can put in (3.49) the total field u, which does not change the result because the incident field does not contribute to the integral.

The far field amplitude Ψ defined by the asymptotic representation (3.8) can be expressed by means of a similar integral as in (3.49). Letting point $\mathbf{R}_0$ tend to infinity and passing to the limit under the integral in (3.49) yields

$$\Psi(\vartheta,\varphi) = \frac{1}{4\pi}\iint\limits_S \left(\Psi_G\left(\mathbf{R}(s),\vartheta,\varphi\right)\frac{\partial u\left(\mathbf{R}(s)\right)}{\partial n}\right.$$

$$\left. - \frac{\partial \Psi_G\left(\mathbf{R}(s),\vartheta,\varphi\right)}{\partial n} u\left(\mathbf{R}(s)\right)\right)\mathrm{d}s. \tag{3.51}$$

Here, angles ϑ and φ in spherical coordinates specify the direction along which the point $\mathbf{R}_0$ tends to infinity, and Ψ_G is the far field amplitude of the point source.

3.7.2 The Far Field Approximation in the Forward Cone

We consider only the directions near the axis of the spheroid. First, we consider the forward directions, that is, assume the angle ϑ to be small, such that $\vartheta\sqrt{kb} = \beta = O(1)$. The main contribution to the integral comes from the middle part of the spheroid, where quick oscillations of Ψ_G and u are self-compensated, while the integrals over the vicinities of the end-points are small. In this middle part of the surface, we substitute the asymptotics of the forward wave derived in Sect. 3.4 in the integral (3.51). Due to the reciprocity principle, Ψ_G coincides with the field of the plane wave incident from the opposite direction, that is from the direction specified by angles $\pi - \vartheta$ and $\pi - \varphi$. Thus, it has the asymptotic approximation expressed by formula (3.17) with all R_n set equal to zero. To change the direction back to angle ϑ, we reflect the z axis, that is replace $\pi - \vartheta$ with ϑ and change the sign of η. This results in the following formula:

$$\Psi_G = \mathrm{e}^{-\mathrm{i}kp\eta}\frac{1}{\sqrt{\chi\tau}\sqrt{1-\eta^2}\beta}\sum_{m=0}^{\infty}\frac{2}{1+\delta_0^m}\cos\left(m[\varphi_0-\varphi]\right)\times$$

$$\times\int\limits_{-\infty}^{+\infty}\left(\frac{1-\eta}{1+\eta}\right)^{is} M_{\mathrm{is},m/2}(\mathrm{i}\beta^2)a_m(s)M_{\mathrm{is},m/2}(-\mathrm{i}\chi\tau)\mathrm{d}s. \tag{3.52}$$

Here, δ_0^m is the Kronecker delta symbol: $\delta_0^m = 1$ for $m = 0$ and $\delta_0^m = 0$ for $m \neq 0$.

It is easy to express the infinitesimal element of the surface ds and the normal derivative presented in the integral (3.51) in coordinates (η, τ) of the boundary layer

$$\mathrm{d}s = a\sqrt{1-\eta^2}\mathrm{d}\varphi\ p\ \mathrm{d}\eta, \qquad \left.\frac{\partial}{\partial n}\right|_{n=0} = \frac{2}{a}\frac{1}{\sqrt{1-\eta^2}}\left.\frac{\partial}{\partial\tau}\right|_{\tau=1}.$$

We substitute these expressions into the integral (3.51) and change the order of integration. In view of the orthogonality of trigonometric functions, only the terms with $n = m$ give a nonzero contribution. Therefore, the double series is reduced to the single one. Further, the integral by η is reduced to delta-function

$$\int_{-1}^{1} \left(\frac{1-\eta}{1+\eta}\right)^{\mathrm{i}(t-s)} \frac{\mathrm{d}\eta}{1-\eta^2} = \pi\delta(t-s), \tag{3.53}$$

which can be easily seen if one changes the integration variable $\eta = -\tanh(\lambda/2)$. This results in the following asymptotic approximation for the far field amplitude:

$$\Psi = \frac{2ib}{\pi\alpha\beta} \sum_{n=0}^{\infty} \frac{1}{(n!)^3} \frac{\cos(n\varphi)}{1+\delta_0^n} \int_{-\infty}^{+\infty} M_{\mathrm{i}t,n/2}(\mathrm{i}\alpha^2) M_{\mathrm{i}t,n/2}(\mathrm{i}\beta^2) \times$$
$$\times R_n(t) \Gamma^2\left(\frac{n+1}{2} + \mathrm{i}t\right) \Gamma\left(\frac{n+1}{2} - \mathrm{i}t\right) \mathrm{d}t. \tag{3.54}$$

The asymptotic approximation for the scattering cross-section [17]

$$\Sigma = \frac{4}{ka^2\sqrt{1+\alpha^2/\chi}} \mathrm{Im}\,\Psi(\vartheta_0, 0), \tag{3.55}$$

which we normalize by the visible cross-section of the spheroid $\pi a^2\sqrt{1+\alpha^2/\chi}$, immediately follows from formula (3.54)

$$\Sigma = -\frac{8}{\pi\chi\alpha^2\sqrt{1+\alpha^2/\chi}} \sum_{n=0}^{\infty} \frac{1}{(n!)^3} \frac{1}{1+\delta_0^n} \times$$
$$\times \int_{-\infty}^{+\infty} \mathrm{Re}\left[M^2_{\mathrm{i}t,n/2}(\mathrm{i}\alpha^2) R_n(t) \Gamma^2\left(\frac{n+1}{2} + \mathrm{i}t\right) \Gamma\left(\frac{n+1}{2} - \mathrm{i}t\right)\right] \mathrm{d}t. \tag{3.56}$$

The integrated expressions in formulas (3.54) and (3.56) rapidly decrease at infinity, however when χ and/or α increase, the interval which contributes to the integrals becomes larger. The effects are very similar to those related to the integrals in the formulas for the currents.

The numerical results for the total scattering cross-section approximation given by (3.56) are plotted for the case of the soft and hard spheroids in Fig. 3.3 versus the elongation parameter χ. We consider the case of axial incidence and the case of incidence at the scaled angles $\alpha = 1$, 2 and 5. The results presented in this figure allow one to conclude that the effect caused by the elongation is in the decrease of the effective cross-section in the case of the hard surface and in its increase in the case of the soft surface. For axial incidence, the effect is maximal and it monotonically decreases when the scaled angle α becomes larger. When $\chi \to +\infty$ and the body becomes not strongly elongated, the effective cross-section Σ tends to its well-known high-frequency limit of 2.

Angular characteristics $|\Psi|$ are presented in Figs. 3.4 and 3.5 for the case of plane wave incidence at 5° on soft and hard spheroids of different aspect ratios. It is worth

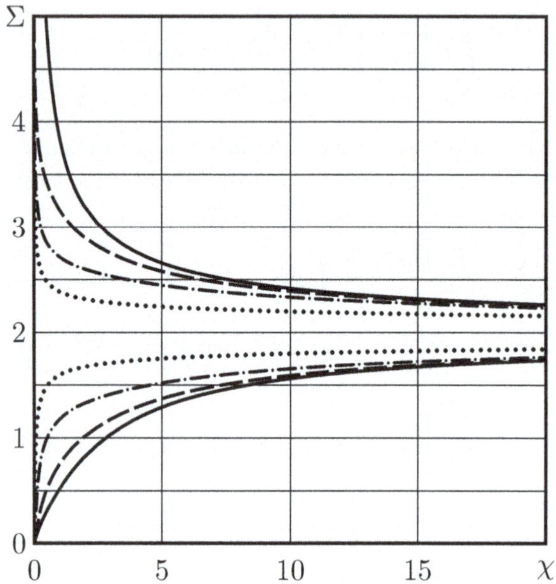

Fig. 3.3 The total scattering cross-section for the elongated soft and hard spheroids as the functions of the elongation parameter χ for axial incidence (solid line), for $\beta_0 = 1$, 2 and 5 (dashed, dash-dotted and dotted lines). The upper curves correspond to the case of the soft surface, and lower ones are for the case of the hard surface

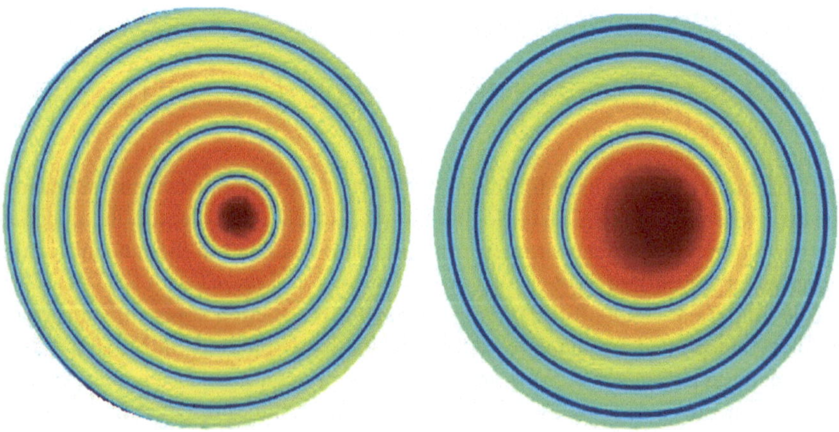

Fig. 3.4 The far field amplitude on the soft spheroids with $kb = 100$, $a : b = 1 : 5$ (left) and $a : b = 1 : 20$ (right) at 5° incidence in a 30° forward cone

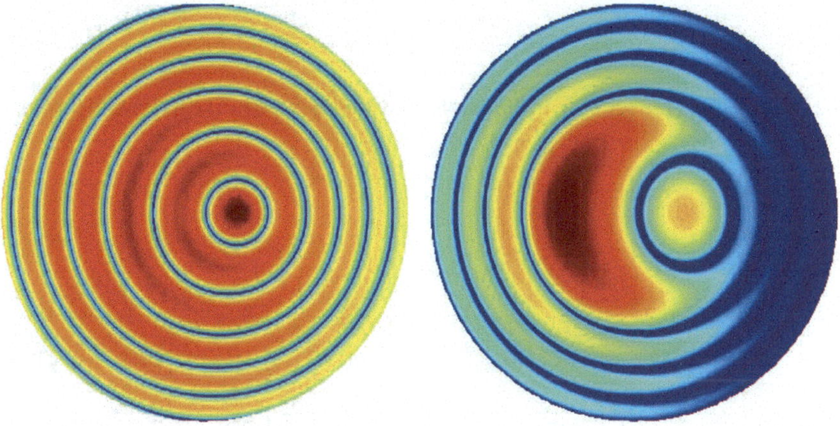

Fig. 3.5 The far field amplitude on the hard spheroids with $kb = 100$, $a : b = 1 : 5$ (left) and $a : b = 1 : 20$ (right) at 5° incidence in a 30° forward cone

noting that usually the main beam of the scattered fields is the shadow beam directed along the incident wave. However, in the case of scattering by hard spheroids with a high aspect ratio, the maximum of the far field amplitude is not in that direction, but in the direction of the ray reflected from the side surface of the body, which is evidently seen in Fig. 3.5.

3.7.3 The Far Field Approximation in the Backward Cone

Formula (3.54) expresses the high-frequency asymptotics of the far field amplitude in the forward directions. In this section, we derive the asymptotic approximations for the far field amplitude in the backward directions, that is for ϑ close to π. The back-scattered field appears as the result of two contributions. One is associated with the reflection of the incident wave from the specular point on the surface of the spheroid (this is the scatterer response from the first Fresnel zone). The other contribution to the field is due to the diffraction effects. The incident wave generates the forward field which propagates along the surface of the spheroid. This wave reaches the shadowed end of the body, encircles it and returns in the form of a backward wave. Detaching from the surface, this backward wave produces the far field in backward cone directions. We denote these two contributions to the far field amplitude as Ψ^{refl} and Ψ^{diffr}.

Consider first the contribution of the specular point. In the case of the small angle of incidence, the specular point lies in a vicinity of the illuminated end-point of the spheroid. In this vicinity, we approximate the surface with the appropriate paraboloid of revolution in the same manner as was done in Sect. 3.5, where the vicinity of the shadowed end was considered. Now the paraboliccoordinates are introduced by the formulas

$$z = -p - \frac{v-w}{2k}, \qquad r = \frac{\sqrt{vw}}{k}. \tag{3.57}$$

The acoustic field reflected from the paraboloid is expressed by the following formula:

$$\begin{aligned} u_n =& \frac{i(-1)^n}{\pi(n!)^2} \frac{\mathrm{e}^{-ikp\cos\vartheta_0}}{\sqrt{\mu\nu}\sin\vartheta_0} \times \\ & \times \int_{-\infty}^{+\infty} \left[\tan\left(\frac{\vartheta_0}{2}\right)\right]^{-2it} \Gamma\left(\frac{n+1}{2}+\mathrm{i}t\right)\Gamma\left(\frac{n+1}{2}-\mathrm{i}t\right) \times \\ & \times M_{\mathrm{i}t,n/2}(-\mathrm{i}w)\left\{M_{\mathrm{i}t,n/2}(-\mathrm{i}v) + R_n(t)W_{\mathrm{i}t,n/2}(-\mathrm{i}v)\right\}\mathrm{d}t. \end{aligned} \tag{3.58}$$

Formula (3.58) is exact. It is derived in [14] (see also [15, Chap. 3]). It is valid for any angle of incidence, but for $\vartheta_0 = 0$ it contains ambiguity of the type zero divided by zero, and one should consider it as the limit. One can note that the structure of formula (3.58) is the same as in expression (3.17). That means that function M in the braces corresponds to the incident wave and the other term expresses the secondary field.

Our immediate goal is to find the far field asymptotics of the above expression. Let

$$z = R\cos\vartheta, \quad r = R\sin\vartheta, \quad R \to +\infty. \tag{3.59}$$

The angle ϑ specifies the direction and for the case of the backscattering considered here, it is close to π. Neglecting smaller order terms, we find

$$v + w = 2kR + kp\cos\vartheta + O(R^{-1}), \tag{3.60}$$

$$\frac{v}{w} = \frac{1-\cos\vartheta}{1+\cos\vartheta} + O(R^{-1}) = \left[\tan\left(\frac{\vartheta}{2}\right)\right]^2 + O(R^{-1}), \tag{3.61}$$

which means that for $\vartheta \neq \pi$ both w and v are large and positive. We replace in (3.58) the Whittaker functions by their asymptotic expansions (3.41) and (A.23). The principal order contribution is given by the first term of (3.41). We get

$$U_n \sim \frac{\mathrm{e}^{\mathrm{i}kR}}{R}\Psi_n^{\mathrm{refl}}, \tag{3.62}$$

where

$$\begin{aligned} \Psi_n^{\mathrm{refl}} =& \frac{\mathrm{e}^{\mathrm{i}kp(\cos\vartheta_0+\cos\vartheta')}}{\pi n!\mathrm{i}^n k\sin\vartheta_0\sin\vartheta'} \times \\ & \times \int_{-(1+\mathrm{i}\varepsilon)\infty}^{+(1+\mathrm{i}\varepsilon)\infty} \Gamma\left(\frac{n+1}{2}+\mathrm{i}t\right) R_n(t)\left[\tan\left(\frac{\vartheta_0}{2}\right)\tan\left(\frac{\vartheta'}{2}\right)\right]^{-2\mathrm{i}t} \mathrm{e}^{\pi t}\mathrm{d}t. \end{aligned}$$

To achieve the convergence of the written integral, one needs to deform the path of integration to the upper half-plane of complex t by introducing a small imaginary part $t \mapsto t + \mathrm{i}\varepsilon$. It is worth noting the symmetry of the above expression with respect to the angles ϑ_0 and $\vartheta' = \pi - \vartheta$, which agrees with the reciprocity principle. For small angles, the integral is reduced to the sum of residues in the poles of the gamma-function

$$\Psi_n^{\mathrm{refl}} = \frac{2\mathrm{i}}{n!} \frac{\mathrm{e}^{-\mathrm{i}kp(\cos\vartheta_0 + \cos\vartheta')}}{k(1+\cos\vartheta_0)(1+\cos\vartheta')} \times \\ \times \sum_{m=0}^{+\infty} \frac{1}{m!} R_n\left(\frac{\mathrm{i}}{2}(1+n+2m)\right) \left[\tan\left(\frac{\vartheta_0}{2}\right)\tan\left(\frac{\vartheta'}{2}\right)\right]^{n+2m}. \tag{3.63}$$

It is nice that this formula does not contain ambiguity for the angles equal to zero. Formula (3.63) describes the reflection of the plane wave from the convex side of the infinite paraboloid of revolution and coincides with the known expressions [11].

Now we consider the contribution brought to the back-scattered field by the backward wave propagating near the surface. To find the far field amplitude corresponding to that field, we perform exactly the same derivations as in Sect. 3.7 with the only difference that instead of the asymptotics of the forward wave we use the asymptotics of the backward wave. This difference results in the additional multiplier $r_n(t)$ in the final formula

$$\Psi_n^{\mathrm{diffr}} = \frac{\mathrm{i}^n b}{(n!)^3} \frac{e^{2\mathrm{i}kp}}{\pi\alpha\beta'} \int_{-\infty}^{+\infty} M_{\mathrm{i}t,n/2}(\mathrm{i}\alpha^2) M_{\mathrm{i}t,n/2}\left(\mathrm{i}(\beta')^2\right) R_n(t) \times \\ \times \Gamma^3\left(\frac{n+1}{2} + \mathrm{i}t\right)(4kb)^{-2\mathrm{i}t}\,\mathrm{d}t.$$

Here we introduced the scaled angle $\beta' = \sqrt{kp}(\pi - \vartheta)$.

For the characterization of the backscattering, we introduce the quantity

$$B = \frac{4kb}{ka^2\sqrt{1+\alpha^2/\chi}} \Psi(\pi - \vartheta_0, \pi). \tag{3.64}$$

Here, compared to formula (3.55) we inserted additional multiplier kb in order to equalize B in view of $\Psi^{\mathrm{refl}} \sim (kb)^{-1}$ at large frequencies. Below we present the values of B in dB.

For soft spheroids, the backscattering amplitude as a function of the wave size is presented in Fig. 3.6. The computations are done for the four spheroids with aspect ratios b/a equal to 5, 10, 20 and 50 (the corresponding dependencies are presented by solid, dashed, dash-dotted and dotted lines), and we consider the case of axial incidence. More thick lines present B^{refl}, and more thin lines present B^{diffr}. These computations allow one to conclude that the main contribution is given by the reflected field. The diffracted field may be significant only in the case of spheroids

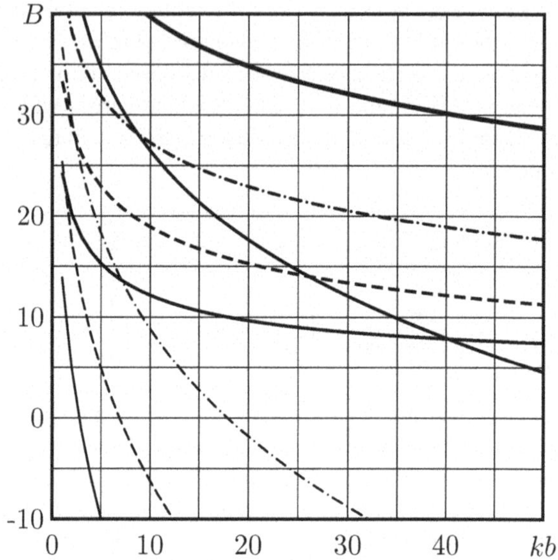

Fig. 3.6 The backscattering cross-section for elongated soft spheroids as a function of the wave size for the case of the axial incidence

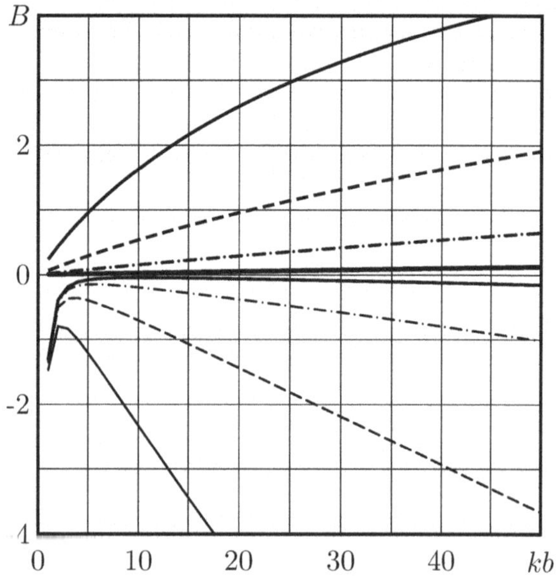

Fig. 3.7 The backscattering cross-section for elongated hard spheroids as a function of wave size for the case of the axial incidence

with not too large wave sizes. Similar results for the case of hard spheroids are presented in Fig. 3.7. The influence of the diffracted component on the total backscattered field in this case is significantly stronger.

The angular characteristics of the backscattering are presented in Fig. 3.8 for soft and hard spheroids with the aspect ratio $b/a = 10$ and different wave sizes kb.

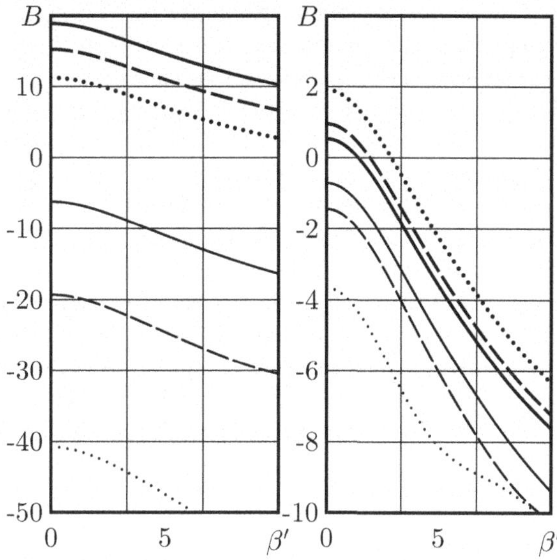

Fig. 3.8 The backscattering cross-section for the elongated soft spheroids (left) and hard spheroids (right) with $b/a = 10$ as a function of the angle. Solid, dashed and dotted lines are for wave sizes $kb = 10$, 20 and 50

3.8 Conclusion and Discussion

In this chapter, we have presented the asymptotic approximation for the field in the problem of acoustic wave diffraction by strongly elongated spheroids. The ideas of derivations and the final expressions for the fields are similar to those used in Chap. 2 except that instead of the two terms we now get the series of Fourier harmonics. In the paraxial regime, this series rapidly converges and for most numerical tests it was sufficient to take into account no more than 10 terms. We point out the main specifics of these asymptotic expressions and the corresponding problems of diffraction. The Fock domain, if defined with the usual classical formulas, contains the whole body. Therefore, the variation of the surface curvature should be taken into account; the curvature cannot be considered constant as in the classical Fock asymptotic expansion. The rate of elongation characterized by parameter χ crucially influences the fields. Though the limiting case of $\chi \to 0$ does not exist, the asymptotic approximations are valid for any arbitrarily small values of χ. For large values of χ, when the body becomes not too much elongated, the asymptotic representations of this chapter reduce the classical Fock approximation. So, though the computational difficulties increase (in particular, the integration interval lengthens and quick oscillations of the integrated expressions appear), the asymptotic expressions of this chapter can be used for arbitrarily large values of χ.

From the point of view of numerical analysis, the problems of this chapter appear more difficult than in the case of diffraction by an elliptic cylinder. For the solvers that do not exploit the symmetry of the surface and deal with 3D formulations, the maximal accessible frequency is limited to $kp \sim 500-1000$, and for the solvers that compute Fourier harmonics it is about 5 times higher. Such frequencies lead to the values of the asymptotic parameter $m_0 = (k\rho_0/2)^{1/3} \leq 25$ in the classical Fock asymptotic expansion, and as shown in Chap. 1 this may result in insufficiently accurate or even wrong results for the currents (see also [10]). One can argue that the Helmholtz equation allows variable separation in spheroidal coordinates, which gives the exact solution in the form of a series involving the spheroidal functions [11, 20]. But, as is usual for the exact solutions of diffraction problems, such representations are not suitable for high frequencies, since the number of terms grows proportionally to the square of frequency. Though there exists (see, for example, [19]) the way to transform the series, which reduces the number of terms to be proportional to the first power of frequency, high frequencies corresponding to $kp \sim 500$–1000 remain difficult to reach using this approach. Thus, the use of the asymptotic approximations derived in this chapter seem to be the only way to carry out a numerical analysis in the case of kp of the order of thousands and $\chi \sim 1$. These expansions give the necessary tool for the validation of the modern solvers, which are being developed in our days.

References

1. M. Abramowitz, I.A. Stegun, Handbook of mathematical functions with formulas, graphs and mathematical tables, National Bureau of Standards. Appl. Math. Ser. **55** (1964)
2. I.V. Andronov, Diffraction by a strongly elongated body of revolution. Acoust. Phys. **57**(2), 121–126 (2011)
3. I.V. Andronov, Diffraction of a plane wave incident at a small angle to the axis of a strongly elongated spheroid. Acoust. Phys. **58**(5), 521–529 (2012)
4. I.V. Andronov, Diffraction of spherical waves on large strongly elongated spheroids. Acta Acustica United Acustica **99**(2), 177–182 (2013)
5. I.V. Andronov, High-Frequency scattering by a strongly elongated body. Acoust. Phys. **59**(4), 369–372 (2013)
6. I.V. Andronov, High-frequency acoustic scattering from prolate spheroids with high aspect ratio. J. Acoustical Soc. Am. **134**(6), 4307–4316 (2013)
7. I.V. Andronov, Point source diffraction by a strongly elongated spheroid. J. Sound Vibr. **355**, 360–368 (2015)
8. I.V. Andronov, Diffraction of a gaussian beam by a strongly elongated spheroid. Acoust. Phys. **65**(4), 335–339 (2018)
9. I.V. Andronov, B.P. Belinskiy, On the parabolic equation method for the problem of diffraction by strongly elongated spheroid. J. Math. Anal. Appl. **456**, 1176–1202 (2017)
10. M.G. Belkina, *Radiation Characteristics of an Elongated Rotary Ellipsoid (in Russian), in Diffraction of Electromagnetic Waves On Certain Bodies of Revolution* (Soviet Radio, Moscow, 1957)
11. J.J. Bowman, T.B.A. Senior, P.L.E. Uslenghi, *Electromagnetic and Acoustic Scattering by Simple Shapes* (North-Holland Publishing Co., Amsterdam, 1969)
12. E.V. Chernokozhin, I.V. Andronov, A. Boag, *Mutual Validation of a Fast Solver Based on the Multilevel Nonuniform Grid Approach and an Asymptotic Approximation for High-frequency Scattering by Strongly Elongated Spheroids, URSI GASS 2020* (Rome, Italy, 2020). https://doi.org/10.23919/URSIGASS49373.2020.9231997
13. E. Chernokozhin, Y. Brick, A. Boag, A fast and stable solver for acoustic scattering problems based on the nonuniform grid approach. J. Acoust. Soc. Am. **139**, 472–480 (2016)
14. V.A. Fock, *Theory of diffraction by a paraboloid of revolution (in Russian), in Diffraction of Electromagnetic Waves by Some Bodies of Revolution* (Soviet Radio, Moscow, 1957), pp.5–56

15. V.A. Fock, *Electromagnetic Diffraction and Propagation Problems* (Pergamon Press, Frankfurt, 1965)
16. I.S. Gradshtein, I.M. Ryzhik, *Table of Integrals, Series and Products* (Elsevier, Academic Press, New York, 2007)
17. H. Hönl, A.W. Maue, K. Westpfahl, Theorie der Beugung, (Springer-Verlag, Berlin). Theory of Diffraction. Naval Intelligence Support Center **1978**, 425 (1961)
18. C.W. Horton, F.C. Karal Jr., On the diffraction of a plane sound wave by a paraboloid of revolution. J. Acoust. Soc. Am. **22**, 855–856 (1950)
19. A.A. Kleschev, Scattering of sound by ideal spheroids in limit case of high frequencies. Sov. Phys. Acoust. **19**, 699–704 (1973)
20. I.V. Komarov, L.I. Ponomarev, *SYu* (Slavyanov, Spheroidal and Coulomb Spheroidal Functions (Science, Moscow, 1976)
21. I.J. Thompson, A.R. Barnett, COULCC: A continued-fraction algorithm for Coulomb functions of complex order with complex arguments. Comput. Phys. Commun. **36**, 363–372 (1985)
22. N.V. Voshchinnikov, V.G. Farafonov, Light scattering by an elongated particle: Spheroid versus infinite cylinder. Meas. Sci. Technol. **13**, 249–255 (2002)

Chapter 4
Electromagnetic Wave Diffraction by Spheroid

4.1 Introduction

In this chapter, we consider the problem of electromagnetic waves diffraction by a perfectly conducting strongly elongated spheroid. It is worth noting that the boundary-value problems for Maxwell equations allow variable separation in spheroidal coordinates only in the case when the fields do not depend on the angular coordinate. Such is the problem of the point dipole field with the dipole being exactly on the axis of the spheroid and the dipole momentum being directed along this axis (In the case of the strongly elongated spheroid, this problem is considered in [8].). In all the other cases, the problems of electromagnetic wave diffraction can be only reduced to infinite systems of linear algebraic equations for the coefficients of the series, which represents the solution in terms of spheroidal functions [16, 19].

The specialized parabolic equation method, which we described in the previous chapters, is applicable to electromagnetic wave diffraction, but requires some minor modifications. In particular, in view of the vector character of electromagnetic fields, one gets for every Fourier harmonics the system of two parabolic equations. However, the leading order approximation for the diffracted field is still given explicitly. As usual, we first find the representation for the field in the boundary layer near the surface and then with the use of the vector version of Green's formula (Stratton-Chu formula [20]) extend the solution to the whole exterior of the spheroid.

The asymptotic analysis presented in this chapter was first applied in [2, 3] and [4] to the case of axially incident plane wave. The more detailed study of this case accompanied by the results of the comparison of the asymptotic approximation with the numerical tests was reported in [10]. The backward wave was studied in [9]. The generalizations to the case of skew incidence were published in [5, 7] and [11]. The far field asymptotics are considered [6, 12, 13] and [14]. The case of dipole source field diffraction is studied in [8].

I. Andronov, *Problems of High Frequency Diffraction by Elongated Bodies*, Springer Series in Optical Sciences 243, https://doi.org/10.1007/978-981-99-1276-6_4

4.2 The Problem Formulation and the Boundary-Layer Coordinates

Let the perfectly conducting spheroid be illuminated by the electromagnetic wave. We remind that we consider stationary wave process and assume the time factor in the form $\mathrm{e}^{-\mathrm{i}\omega t}$. The geometry of the problem is shown in Fig. 4.1. The electromagnetic field satisfies Maxwell's equations

$$\begin{cases} \nabla \times \mathbf{E} = \mathrm{i}k\mathbf{H}, \\ \nabla \times \mathbf{H} = -\mathrm{i}k\mathbf{E}, \end{cases} \tag{4.1}$$

which we write in the symmetric form by introducing vectors **E** and **H**, where **E** is the electric vector divided by the characteristic impedance $\sqrt{\mu/\varepsilon}$ of the space and **H** is the magnetic vector. The wave number is $k = \omega\sqrt{\varepsilon\mu}$. The boundary condition on the surface expresses the absence of tangential components of the electric field and can be written as

$$\mathbf{E} - (\mathbf{E}, \mathbf{e}_n)\mathbf{e}_n = 0, \tag{4.2}$$

where $\mathbf{e}_n$ is the normal to the surface. The scattered field $\mathbf{E}^s$, $\mathbf{H}^s$ which is the difference between the total field and the incident field $\mathbf{E}^{\mathrm{inc}}$, $\mathbf{H}^{\mathrm{inc}}$, satisfies the Sommerfeld radiation conditions, i.e. at infinitely large distances it represents an outgoing wave process.

We assume that the size of the body is large, that is, both ka and kb are large quantities. The asymptotic parameter is introduced as kp, where $p = \sqrt{b^2 - a^2}$ is the half focal distance of the spheroid. We consider the case of a strongly elongated body and characterize the rate of elongation by the usual parameter

$$\chi = \frac{ka^2}{p},$$

which we assume of the order $O(1)$. Besides, we consider paraxial regime, that is, the direction of incidence is assumed to form a small angle ϑ_0 with the axis of

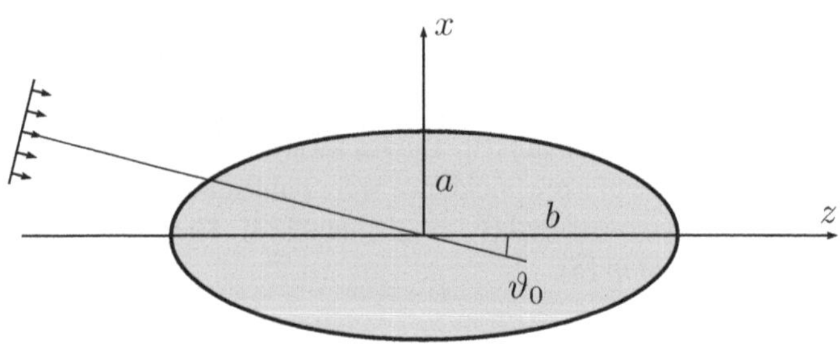

Fig. 4.1 Geometry of the problem

the spheroid. More precisely, we consider the scaled angle to be bounded, that is, $\alpha = \sqrt{kp}\vartheta_0 \le const = O(1)$.

The two cases of the incident field are considered: the case of the plane wave incidence, and the case of the field radiated by a point dipole.

The field in the boundary layer near the surface can be represented as the sum of the waves that propagate in the forward and backward directions. The principal forward wave is excited by the incident field. This forward wave reaches the rear ending of the spheroid. There, it partly irradiates and partly encircles the rear tip and forms the backward wave. The backward wave travels along the surface in the opposite direction, it reaches the front tip and after encircling it forms the secondary forward wave, which in turn forms the secondary backward wave, etc. In the problems considered in Chaps. 2 and 3, it was sufficient to take into account only the principal forward wave and the principal backward wave. Different from the acoustic problems of diffraction in the electromagnetic case, the contribution of the secondary forward wave may be noticeable (see Sect. 4.7).

In the boundary layer near the surface, we identify three subdomains. In the middle part, we use spheroidal coordinates (η, ξ, φ)

$$\begin{aligned} x &= p\sqrt{\xi^2 - 1}\sqrt{1 - \eta^2}\sin\varphi, \\ y &= p\sqrt{\xi^2 - 1}\sqrt{1 - \eta^2}\cos\varphi, \\ z &= p\xi\eta, \end{aligned} \tag{4.3}$$

in which we scale the radial coordinate ξ by replacing it with τ introduced by the formula

$$\xi = 1 + \frac{2\chi\tau}{kp}.$$

Then up to smaller order terms $\tau = 1$ on the surface.

At the vicinities of the spheroid endings, we approximate the surface with the surfaces of the appropriately chosen paraboloids. This enables us to separate the variables in Maxwell equations (without introducing the parabolic equation approximation) and use the exact expression for the field derived by V. A. Fock in [17]. The paraboloidal coordinates (v, w, φ) associated with the rear tip are introduced by the formulas

$$x = \frac{\sqrt{vw}}{k}\sin\varphi, \qquad y = \frac{\sqrt{vw}}{k}\cos\varphi \qquad z = p + \frac{v - w}{2k}. \tag{4.4}$$

The paraboloid, which approximates the spheroidal surface near the tip, is given by the equation $v = \chi$.

The coordinates w and η are related to each other by the formula

$$\eta = 1 - \frac{w}{2kp} + \cdots. \tag{4.5}$$

4.3 Representation for the Forward Wave

Having in mind the structure of the fields in the boundary layer near the surface, we start our analysis by finding the asymptotic representation for the principal forward wave. We use the same approach as in Chaps. 2 and 3. It is based on the parabolic equation method in coordinates (η, τ, φ). Maxwell equations (4.1) in spheroidal coordinates (4.3) take the form

$$\frac{\partial}{\partial\eta}h_\varphi E_\varphi - \frac{\partial}{\partial\varphi}h_\eta E_\eta = -ikh_\eta h_\varphi H_\xi, \tag{4.6}$$

$$\frac{\partial}{\partial\varphi}h_\xi E_\xi - \frac{\partial}{\partial\xi}h_\varphi E_\varphi = -ikh_\varphi h_\xi H_\eta,$$

$$\frac{\partial}{\partial\xi}h_\eta E_\eta - \frac{\partial}{\partial\eta}h_\xi E_\xi = -ikh_\xi h_\eta H_\varphi,$$

$$\frac{\partial}{\partial\eta}h_\varphi H_\varphi - \frac{\partial}{\partial\varphi}h_\eta H_\eta = ikh_\eta h_\varphi E_\xi,$$

$$\frac{\partial}{\partial\varphi}h_\xi H_\xi - \frac{\partial}{\partial\xi}h_\varphi H_\varphi = ikh_\varphi h_\xi E_\eta,$$

$$\frac{\partial}{\partial\xi}h_\eta H_\eta - \frac{\partial}{\partial\eta}h_\xi H_\xi = ikh_\xi h_\eta E_\varphi, \tag{4.7}$$

where h_ξ, h_η and h_φ are the metric coefficients

$$h_\xi = p\sqrt{\frac{\xi^2-\eta^2}{\xi^2-1}} \approx p\sqrt{kp}\frac{\sqrt{1-\eta^2}}{\sqrt{\chi\tau}}, \quad h_\eta = p\sqrt{\frac{\xi^2-\eta^2}{1-\eta^2}} \approx p,$$

$$h_\varphi = p\sqrt{\xi^2-1}\sqrt{1-\eta^2} \approx \frac{p}{\sqrt{kp}}\sqrt{\chi\tau(1-\eta^2)}.$$

In view of the symmetry of the revolution of the surface, it is convenient to represent the field as the sum of Fourier harmonics

$$\mathbf{E} = \sum_{n=-\infty}^{+\infty} \mathbf{E}_n \mathrm{e}^{\mathrm{i}n\varphi}, \qquad \mathbf{H} = \sum_{n=-\infty}^{+\infty} \mathbf{H}_n \mathrm{e}^{\mathrm{i}n\varphi}. \tag{4.8}$$

Then each of the harmonics satisfies a separate boundary-value problem.

The components E_ξ, E_η, H_ξ and H_η of the field of each harmonics $\mathbf{E}_n$, $\mathbf{H}_n$ can be express with the help of equation numbers one, two, four and five of system (4.6)–(4.7) via only two unknown functions $E_n \equiv E_{n,\varphi}$ and $H_n \equiv H_{n,\varphi}$. In particular,

$$E_{n,\eta} = \frac{\mathrm{i}}{(k^2h_\varphi^2 - n^2)h_\eta h_\xi}\left\{\frac{2k^2p}{\chi}h_\eta h_\varphi\frac{\partial}{\partial\tau}\left(h_\varphi H_n\right) + nh_\xi\frac{\partial}{\partial\eta}\left(h_\varphi E_n\right)\right\}$$

$$\approx \frac{2\mathrm{i}}{\sqrt{kp\chi(1-\eta^2)}}\frac{\partial}{\partial\tau}\left(\sqrt{\tau}H_n\right).$$

Following the parabolic equation method, we extract the quick oscillating factor

$$E_n = \mathrm{e}^{\mathrm{i}kp\eta}\mathcal{E}_n(\eta,\tau), \qquad H_n = \mathrm{e}^{\mathrm{i}kp\eta}\mathcal{H}_n(\eta,\tau) \tag{4.9}$$

and assume that the attenuation functions $\mathcal{E}_n$ and $\mathcal{H}_n$ depend on their arguments in such a way that the derivatives with respect to η and τ do not change the asymptotic order. Generally, the attenuation functions can be searched in the form of asymptotic series with respect to the inverse powers of kp. However, we restrict our analysis to finding only the leading order approximation. So, substituting representations (4.9) into Maxwell's equations (4.6)–(4.7) and neglecting smaller order terms, we get the system of parabolic equations for the attenuation functions $\mathcal{E}_n$ and $\mathcal{H}_n$

$$\begin{cases} L_{1+n^2}\mathcal{E}_n + \frac{\mathrm{i}n}{2\tau}\mathcal{H}_n = 0, \\ L_{1+n^2}\mathcal{H}_n - \frac{\mathrm{i}n}{2\tau}\mathcal{E}_n = 0, \end{cases}$$

where

$$L_s = \tau\frac{\partial^2}{\partial\tau^2} + \frac{\partial}{\partial\tau} + \mathrm{i}\chi\frac{1-\eta^2}{2}\frac{\partial}{\partial\eta} + \left(\frac{\chi^2}{4}\tau - \frac{s}{4\tau} - \frac{\mathrm{i}\chi\eta}{2}\right). \tag{4.10}$$

This system splits into two independent parabolic equations

$$L_{(n-1)^2}A_n = 0, \qquad L_{(n+1)^2}B_n = 0 \tag{4.11}$$

for the new unknowns A_n and B_n which we introduce by the formulas

$$\mathcal{E}_n = A_n + B_n, \qquad \mathcal{H}_n = -\mathrm{i}(A_n - B_n). \tag{4.12}$$

The boundary conditions simplify at the leading order to

$$A_n + B_n = 0, \qquad \left(\frac{1}{2} + \frac{\partial}{\partial\tau}\right)\left(A_n - B_n\right) = 0. \tag{4.13}$$

It is convenient to exclude negative indices in the series (4.8). For that, we decompose the electromagnetic field into TE and TM components. Consider first the case of TE polarization. In that case, $E_{-n} = E_n$ and $H_{-n} = -H_n$. Then changing the notations $2E_n \mapsto E_n$ and $2\mathrm{i}H_n \mapsto H_n$, we get

$$E_\varphi = \mathrm{e}^{ikp\eta} \sum_{n=0}^{+\infty} \mathcal{E}_n \cos(n\varphi), \qquad H_\varphi = \mathrm{e}^{ikp\eta} \sum_{n=0}^{+\infty} \mathcal{H}_n \sin(n\varphi). \tag{4.14}$$

For every n, the attenuation functions $\mathcal{E}_n$ and $\mathcal{H}_n$ are subject to the system of parabolic equations, which reduces to (4.11). Equations (4.11) allow variable separation which enables us to represent the solutions in the form of the integral with respect to the parameter t of variable separation. Representing the fields of the incident wave and the forward wave separately, we get

$$A_n^{\text{forward}} = \frac{1}{\sqrt{\chi\tau}\sqrt{1-\eta^2}} \int_{-\infty}^{+\infty} \left(\frac{1-\eta}{1+\eta}\right)^{it} a_n^{\text{forward}}(t) W_{it,\frac{n-1}{2}}(-\mathrm{i}\chi\tau)\mathrm{d}t, \tag{4.15}$$

$$B_n^{\text{forward}} = \frac{1}{\sqrt{\chi\tau}\sqrt{1-\eta^2}} \int_{-\infty}^{+\infty} \left(\frac{1-\eta}{1+\eta}\right)^{it} b_n^{\text{forward}}(t) W_{it,\frac{n+1}{2}}(-\mathrm{i}\chi\tau)\mathrm{d}t. \tag{4.16}$$

$$A_n^{\text{inc}} = \frac{1}{\sqrt{\chi\tau}\sqrt{1-\eta^2}} \int_{-\infty}^{+\infty} \left(\frac{1-\eta}{1+\eta}\right)^{it} a_n^{\text{inc}}(t) M_{it,\frac{n-1}{2}}(-\mathrm{i}\chi\tau)\mathrm{d}t, \tag{4.17}$$

$$B_n^{\text{inc}} = \frac{1}{\sqrt{\chi\tau}\sqrt{1-\eta^2}} \int_{-\infty}^{+\infty} \left(\frac{1-\eta}{1+\eta}\right)^{it} b_n^{\text{inc}}(t) M_{it,\frac{n+1}{2}}(-\mathrm{i}\chi\tau)\mathrm{d}t. \tag{4.18}$$

Equations (4.15) and (4.17) are valid for $n = 1, 2, 3, \ldots$, while $A_0^{\text{forward}} = A_0^{\text{inc}} = 0$. Equations (4.16) and (4.18) hold for $n = 0, 1, 2, \ldots$.

The Whittaker functions W are chosen in (4.15) and (4.16) in agreement with the radiation condition. In the corresponding formulas for the incident wave, they are replaced with Whittaker functions M, which are regular at $\tau = 0$, i.e. on the axis of the spheroid.

The amplitudes a_n^{inc} and b_n^{inc} are determined when matching with the incident wave, which will be done in the next section. The amplitudes $a_n^{\text{forward}}(t)$ and $b_n^{\text{forward}}(t)$ in the representation of the forward wave should be chosen such that the boundary conditions (4.13) are satisfied. For the harmonics of the zeroth order, since $H_0 \equiv 0$, the boundary conditions reduce to $\mathcal{E}_0 = 0$. Thus,

$$b_0^{\text{forward}} = -b_0^{\text{inc}} \frac{M_{it,\frac{1}{2}}(-\mathrm{i}\chi)}{W_{it,\frac{1}{2}}(-\mathrm{i}\chi)}. \tag{4.19}$$

The boundary conditions for the harmonics of orders $n = 1, 2, 3, \ldots$ yield the system for the amplitudes

$$\begin{cases} a_n^{\text{forward}} W_{\mathrm{i}t,\frac{n-1}{2}}(-\mathrm{i}\chi) + b_n^{\text{forward}} W_{\mathrm{i}t,\frac{n+1}{2}}(-\mathrm{i}\chi) = \\ \qquad\qquad - a_s^{\text{inc}} M_{\mathrm{i}t,\frac{n-1}{2}}(-\mathrm{i}\chi) - b_n^{\text{inc}} M_{\mathrm{i}t,\frac{n+1}{2}}(-\mathrm{i}\chi), \\ a_n^{\text{forward}} W'_{\mathrm{i}t,\frac{n-1}{2}}(-\mathrm{i}\chi) - b_n^{\text{forward}} W'_{\mathrm{i}t,\frac{n+1}{2}}(-\mathrm{i}\chi) = \\ \qquad\qquad - a_n^{\text{inc}} M'_{\mathrm{i}t,\frac{n-1}{2}}(-\mathrm{i}\chi) + b_n^{\text{inc}} M'_{\mathrm{i}t,\frac{n+1}{2}}(-\mathrm{i}\chi), \end{cases} \tag{4.20}$$

where prime denotes the derivative of the function.

In view of the formula

$$\dot{M}_{\mathrm{i}t,\ell}(z) W_{\mathrm{i}t,\ell}(z) - M_{\mathrm{i}t,\ell}(z) \dot{W}_{\mathrm{i}t,\ell}(z) = \frac{\Gamma(2\ell+1)}{\Gamma\left(\ell + \frac{1}{2} - \mathrm{i}t\right)} \tag{4.21}$$

for the Wronskian of Whittaker functions, one gets from (4.20) the following expressions for the amplitudes:

$$a_n^{\text{forward}} = -\frac{a_n^{\text{inc}}}{Z_n}\left(M_{\mathrm{i}t,\frac{n-1}{2}}(-\mathrm{i}\chi) W'_{\mathrm{i}t,\frac{n+1}{2}}(-\mathrm{i}\chi) + M'_{\mathrm{i}t,\frac{n-1}{2}}(-\mathrm{i}\chi) W_{\mathrm{i}t,\frac{n+1}{2}}(-\mathrm{i}\chi)\right) + \frac{b_n^{\text{inc}}}{Z_n} \frac{\Gamma(n+2)}{\Gamma\left(\frac{n}{2}+1-\mathrm{i}t\right)}, \tag{4.22}$$

$$b_n^{\text{forward}} = \frac{a_n^{\text{inc}}}{Z_n} \frac{\Gamma(n)}{\Gamma\left(\frac{n}{2} - \mathrm{i}t\right)} - \frac{b_n^{\text{inc}}}{Z_n}\left(M_{\mathrm{i}t,\frac{n+1}{2}}(-\mathrm{i}\chi) W'_{\mathrm{i}t,\frac{n-1}{2}}(-\mathrm{i}\chi) + M'_{\mathrm{i}t,\frac{n+1}{2}}(-\mathrm{i}\chi) W_{\mathrm{i}t,\frac{n-1}{2}}(-\mathrm{i}\chi)\right). \tag{4.23}$$

Here

$$Z_n = W_{\mathrm{i}t,\frac{n+1}{2}}(-\mathrm{i}\chi) W'_{\mathrm{i}t,\frac{n-1}{2}}(-\mathrm{i}\chi) + W'_{\mathrm{i}t,\frac{n+1}{2}}(-\mathrm{i}\chi) W_{\mathrm{i}t,\frac{n-1}{2}}(-\mathrm{i}\chi). \tag{4.24}$$

The denominator Z_n can be rewritten in a different form. For that, we use formulas (A.18) and (A.19) to express the derivative of the Whittaker functions in (4.24) which yields

$$Z_n = \left(\frac{it}{n} - \frac{1}{2}\right) W_{\mathrm{i}t,\frac{n+1}{2}}(-\mathrm{i}\chi)^2 - \left(\frac{it}{n} + \frac{1}{2}\right) W_{\mathrm{i}t,\frac{n-1}{2}}(-\mathrm{i}\chi)^2. \tag{4.25}$$

Using the formula (4.21), we can express $M'_{\mathrm{i}t,\frac{n\pm1}{2}}(-\mathrm{i}\chi)$ in (4.22) and (4.23) via other functions, which yields an alternative expression for the amplitudes

$$a_n^{\text{forward}} = -a_n^{\text{inc}} \frac{M_{\text{i}t,\frac{n-1}{2}}(-\text{i}\chi)}{W_{\text{i}t,\frac{n-1}{2}}(-\text{i}\chi)} + \frac{a_n^{\text{inc}}}{Z_n} \frac{\Gamma(n)}{\Gamma\left(\frac{n}{2}-\text{i}t\right)} \frac{W_{\text{i}t,\frac{n+1}{2}}(-\text{i}\chi)}{W_{\text{i}t,\frac{n-1}{2}}(-\text{i}\chi)} + \frac{b_n^{\text{inc}}}{Z_n} \frac{\Gamma(n+2)}{\Gamma\left(\frac{n}{2}+1-\text{i}t\right)} \frac{W_{\text{i}t,\frac{n-1}{2}}(-\text{i}\chi)}{W_{\text{i}t,\frac{n+1}{2}}(-\text{i}\chi)}, \tag{4.26}$$

$$b_n^{\text{forward}} = -b_n^{\text{inc}} \frac{M_{\text{i}t,\frac{n+1}{2}}(-\text{i}\chi)}{W_{\text{i}t,\frac{n+1}{2}}(-\text{i}\chi)} + \frac{a_n^{\text{inc}}}{Z_n} \frac{\Gamma(n)}{\Gamma\left(\frac{n}{2}-\text{i}t\right)} \frac{W_{\text{i}t,\frac{n+1}{2}}(-\text{i}\chi)}{W_{\text{i}t,\frac{n-1}{2}}(-\text{i}\chi)} - \frac{b_n^{\text{inc}}}{Z_n} \frac{\Gamma(n+2)}{\Gamma\left(\frac{n}{2}+1-\text{i}t\right)} \frac{W_{\text{i}t,\frac{n-1}{2}}(-\text{i}\chi)}{W_{\text{i}t,\frac{n+1}{2}}(-\text{i}\chi)}. \tag{4.27}$$

In the case of TM polarization, representation (4.14) changes to

$$E_\varphi = \sum_{n=1}^{+\infty} E_n \sin(n\varphi), \qquad H_\varphi = \sum_{n=0}^{+\infty} H_n \cos(n\varphi). \tag{4.28}$$

Now $E_0 \equiv 0$ and instead of (4.19), one gets

$$b_0^{\text{forward}} = -b_0^{\text{inc}} \frac{M'_{\text{i}t,\frac{1}{2}}(-\text{i}\chi)}{W'_{\text{i}t,\frac{1}{2}}(-\text{i}\chi)}. \tag{4.29}$$

All the other amplitudes are as previously subject to the system (4.20) and therefore are given by formulas (4.22) and (4.23) or equivalently by formulas (4.26) and (4.27).

4.4 Matching with the Incident Wave

In this section, we determine the amplitudes a_n^{inc} and b_n^{inc} in the representation (4.17) and (4.18) for the incident wave. We consider two cases: plane wave incidence and the field of a point dipole source.

4.4.1 The Plane Wave Incidence

Arbitrarily polarized plane wave can be represented as the sum of TE and TM waves. We shall consider these waves separately. The field of the TE plane wave has the following electric and magnetic vectors

$$\mathbf{E}^{TE} = \exp((\text{i}kz\cos\vartheta_0 + \text{i}kx\sin\vartheta_0)\,\mathbf{e}_y,$$

$$\mathbf{H}^{TE} = \exp((ikz\cos\vartheta_0 + ikx\sin\vartheta_0)\{-\cos\vartheta_0\mathbf{e}_x + \sin\vartheta_0\mathbf{e}_z\},$$

where $\mathbf{e}_x$, $\mathbf{e}_y$ and $\mathbf{e}_z$ are unit vectors in Cartesian coordinates.

In cylindrical coordinates (r, z, φ), the transverse components of these vectors are expressed via Bessel functions in the form

$$E_\varphi^{TE} = \mathrm{e}^{ikz\cos\vartheta_0}\Big\{\mathrm{i}J_1(kr\sin\vartheta_0) + \sum_{n=1}^{\infty}\mathrm{i}^{n-1}\Big(J_{n-1}(kr\sin\vartheta_0) - J_{n+1}(kr\sin\vartheta_0)\Big)\cos(n\varphi)\Big\},$$

$$H_\varphi^{TE} = \mathrm{e}^{ikz\cos\vartheta_0}\cos\vartheta_0\sum_{n=1}^{\infty}\mathrm{i}^{n-1}\Big(J_{n-1}(kr\sin\vartheta_0) + J_{n+1}(kr\sin\vartheta_0)\Big)\sin(n\varphi).$$

In order to rewrite these formulas in the form of representation (4.17) and (4.18), it is necessary to decompose the argument of the exponential into the series with respect to the inverse powers of the large parameter $\sqrt{kb}$ and neglect the vanishing at $kb \to \infty$ terms,

$$kz\cos\vartheta_0 \approx kb\eta + \frac{\eta}{2}\left(\chi(\tau-1) - \alpha^2\right), \qquad \alpha = \sqrt{kb}\,\vartheta_0.$$

We remind that the incident angle ϑ_0 is assumed to be small, so that $\alpha \le const = O(1)$.

Each Fourier component of the incident field is a solution of the (4.12) and, therefore, can be represented in the form of (4.17) and (4.18). In order to find the amplitudes a_n^{inc} and b_n^{inc} in these representations, one can equate both formulas for the solution at some fixed value of coordinate τ. This gives the system of integral equations. The left-hand sides of these equations can be identified with the transform (A.32). Therefore, the solution of these equations can be written explicitly. These calculations were done in the previous chapter (see formulas (3.23)–(3.28)) and we can use the result

$$\exp\left(\mathrm{i}\frac{\eta}{2}\left(\chi\tau - \alpha^2\right)\right) J_\ell\left(\sqrt{1-\eta^2}\sqrt{\chi\tau}\alpha\right) = \\ = \frac{1}{\pi}\frac{1}{\sqrt{1-\eta^2}\sqrt{\chi\tau}\alpha}\int_{-\infty}^{+\infty}\left(\frac{1-\eta}{1+\eta}\right)^{\mathrm{i}t} M_{\mathrm{i}t,\ell/2}(\mathrm{i}\alpha^2)\Omega_\ell(t)M_{\mathrm{i}t,\ell/2}(-\mathrm{i}\chi\tau)\mathrm{d}t,$$

where

$$\Omega_\ell = \frac{\Gamma(\ell/2 + 1/2 + \mathrm{i}t)\Gamma(\ell/2 + 1/2 - \mathrm{i}t)}{\Gamma^2(\ell+1)}. \tag{4.30}$$

So, the integral representation for the incident plane wave in the case of TE polarization takes the form

$$E_{\varphi}^{\mathrm{inc}} = \frac{e^{ikp\eta}}{\pi\sqrt{1-\eta^2}\sqrt{\chi\tau}\alpha} \int\limits_{-\infty}^{+\infty} \left(\frac{1-\eta}{1+\eta}\right)^{it} \Bigg\{ \mathrm{i}\Omega_1(t) M_{\mathrm{i}t,1/2}(\mathrm{i}\alpha^2) M_{\mathrm{i}t,1/2}(-\mathrm{i}\chi\tau)$$

$$+ \sum_{n=1}^{\infty} \mathrm{i}^{n-1} \Big(\Omega_{n-1}(t) M_{\mathrm{i}t,(n-1)/2}(\mathrm{i}\alpha^2) M_{\mathrm{i}t,(n-1)/2}(-\mathrm{i}\chi\tau)$$

$$- \Omega_{n+1}(\lambda) M_{\mathrm{i}t,(n+1)/2}(\mathrm{i}\alpha^2) M_{\mathrm{i}t,(n+1)/2}(-\mathrm{i}\chi\tau) \Big) \cos(n\varphi) \Bigg\} \mathrm{d}t,$$

$$H_{\varphi}^{\mathrm{inc}} = \frac{e^{ikp\eta}}{\pi\sqrt{1-\eta^2}\sqrt{\chi\tau}\alpha} \int\limits_{-\infty}^{+\infty} \left(\frac{1-\eta}{1+\eta}\right)^{it} \times$$

$$\times \Bigg\{ \sum_{n=1}^{\infty} \mathrm{i}^{n-1} \Big(\Omega_{n-1}(t) M_{\mathrm{i}t,(n-1)/2}(\mathrm{i}\alpha^2) M_{\mathrm{i}t,(n-1)/2}(-\mathrm{i}\chi\tau)$$

$$+ \Omega_{n+1}(t) M_{\mathrm{i}t,(n+1)/2}(\mathrm{i}\alpha^2) M_{\mathrm{i}t,(n+1)/2}(-\mathrm{i}\chi\tau) \Big) \sin(n\varphi) \Bigg\} \mathrm{d}t.$$

Comparing the above formulas with the representation (4.17) and (4.18) for the auxiliary functions A_n^{inc} and B_n^{inc} corresponding to the incident field, it is easy to find the amplitudes

$$\begin{aligned} a_n^{\mathrm{inc}} &= \frac{\mathrm{i}^{n-1}}{\pi\alpha} \Omega_{n-1} M_{\mathrm{i}t,\frac{n-1}{2}}(\mathrm{i}\alpha^2), \\ b_n^{\mathrm{inc}} &= \frac{\mathrm{i}^{n+1}}{\pi\alpha} \Omega_{n+1} M_{\mathrm{i}t,\frac{n+1}{2}}(\mathrm{i}\alpha^2). \end{aligned} \tag{4.31}$$

The case of TM polarization can be considered similarly. The final expressions for the amplitudes are the same as (4.31) except that a_n^{inc} has the opposite sign.

4.4.2 Point Dipole Source Field Incidence

Now we consider the case when the field excited by a point dipole source is incident on the spheroid. Let the dipole be in the point with coordinates $(X_0, 0, Z_0)$ and its dipole momentum be **D**. First, we describe the field of the point dipole source in local coordinates associated with the dipole. Then, when transforming from the local coordinates to the global ones, we have to consider three cases of the dipole momentum orientation.

4.4.2.1 Representation of the Dipole Field in Local Coordinates

We introduce local spherical coordinates (r, θ, φ) with the center at the point of dipole location and axis $\theta = 0$ coinciding with the direction of the dipole momentum. Then the field radiated by the dipole can be written in the well-known form as

$$\mathbf{H}^{\text{inc}} = -\frac{k^2 D}{4\pi}\frac{\mathrm{e}^{ikr}}{r}\left(1 + \frac{\mathrm{i}}{kr}\right)\sin\theta\,\mathbf{e}_{\varphi}. \tag{4.32}$$

Here D is the dipole momentum magnitude ($D = \|\mathbf{D}\|$). Using Maxwell equations (4.1), it is a simple matter to get the expression for the electric field:

$$\mathbf{E}^{\text{inc}} = \frac{k^2 D}{4\pi}\frac{\mathrm{e}^{ikr}}{r}\left[\left(\frac{2\mathrm{i}}{kr} - \frac{2}{k^2r^2}\right)\cos\theta\,\mathbf{e}_r + \left(1 + \frac{\mathrm{i}}{kr} - \frac{1}{k^2r^2}\right)\sin\theta\,\mathbf{e}_{\theta}\right]. \tag{4.33}$$

Formulas (4.32) and (4.33) are exact, i.e. contain no approximation. However, we need the approximate expressions for these fields, which can be represented in the form of the integrals (4.17) and (4.18) written in the coordinate system of the spheroid. Such an approximation is only possible if the rays launched by the dipole impinge the spheroid at a small angle to its axis, i.e. Cartesian coordinates axis Oz. This condition is equivalent to the assumption that the coordinates $(X_0, 0, Z_0)$ of the dipole satisfy the inequalities

$$Z_0 < -b, \qquad |X_0| \le const = O(a).$$

Further, it is convenient to introduce dimensionless coordinates $\zeta_0 = -Z_0/b$, $\gamma_0 = X_0/a$, and, up to asymptotically small terms, represent the exponential factor as

$$kr \approx kp\eta - kZ_0 + \frac{\chi\tau\eta}{2} + \Phi - \Psi\cos\varphi,$$

where

$$\Phi = \frac{\chi}{2}\frac{\gamma_0^2 + \tau(1-\eta^2)}{\eta + \zeta_0}, \qquad \Psi = \frac{\chi\gamma_0\sqrt{\tau}\sqrt{1-\eta^2}}{\eta + \zeta_0}. \tag{4.34}$$

The dipole oriented along OZ axis

In the case of the dipole with its momentum oriented along the axis of the spheroid, the local coordinate system associated with the dipole and the global coordinates system associated with the spheroid differ only by some shift of coordinates. At the leading order with respect to kp, one can easily derive the following approximation for the incident field:

$$
{}^{z}H_{\varphi} = \frac{k^2 D_z}{4\pi p} \frac{\sqrt{\chi}}{\sqrt{kp}} \frac{\gamma_0 \cos\varphi - \sqrt{1-\eta^2}\sqrt{\tau}}{(\eta+\zeta_0)^2} \times
\\
\times \exp\left\{ikp\eta - ikZ_0 + \mathrm{i}\frac{\chi\tau\eta}{2} + \mathrm{i}\Phi - \mathrm{i}\Psi\cos\varphi\right\},
$$

$$
{}^{z}E_{\varphi} = \frac{k^2 D_z}{4\pi p} \frac{\sqrt{\chi}}{\sqrt{kp}} \frac{\gamma_0 \sin\varphi}{(\eta+\zeta_0)^2} \exp\left\{ikp\eta - ikZ_0 + \mathrm{i}\frac{\chi\tau\eta}{2} + \mathrm{i}\Phi - \mathrm{i}\Psi\cos\varphi\right\}.
$$

Here and below, the superscript ^{z} marks the quantities related to the dipole orientation along the Oz axis.

Computing Fourier coefficients of the incident field in the representation (4.17) and (4.18), we notice the incident wave has TM polarization and

$$
{}^{z}A_n^{\mathrm{inc}} = \Omega_z(-\mathrm{i})^{n-1} f_n(\eta,\tau), \qquad {}^{z}B_n^{\mathrm{inc}} = \Omega_z(-\mathrm{i})^{n-1} g_n(\eta,\tau), \tag{4.35}
$$

where

$$
\Omega_z = \frac{k^2 D_z \sqrt{\chi}}{4p\sqrt{kp}} \mathrm{e}^{-ikZ_0},
$$

$$
f_n(\eta,\tau) = \frac{1}{(\eta+\zeta_0)^2} \exp\left(\frac{\mathrm{i}\chi}{2}\tau\eta + \mathrm{i}\Phi\right) \times
\\
\times \left[\gamma_0 J_{n-1}(\Psi) + \mathrm{i}\sqrt{\tau}\sqrt{1-\eta^2} J_n(\Psi)\right], \tag{4.36}
$$

$$
g_n(\eta,\tau) = \frac{1}{(\eta+\zeta_0)^2} \exp\left(\frac{\mathrm{i}\chi}{2}\tau\eta + \mathrm{i}\Phi\right) \times
\\
\times \left[\gamma_0 J_{n+1}(\Psi) - \mathrm{i}\sqrt{\tau}\sqrt{1-\eta^2} J_n(\Psi)\right].
$$

Here, $J_n()$ are the Bessel functions, and quantities Φ and Ψ are defined by (4.34).

One can check by direct substitution that functions expressed by formulas (4.35) satisfy the parabolic equations

$$
L_{n-1}{}^{z}A_n^{\mathrm{inc}} = 0, \qquad L_{n+1}{}^{z}B_n^{\mathrm{inc}} = 0.
$$

So, these fields can be represented in the form of the integrals (4.17) and (4.18) with appropriate functions a_n^{inc} and b_n^{inc}. This will be done in Sect. 4.4.2.2.

Dipoles oriented along axes OX and OY

Consider now the case when the dipole momentum is oriented along the axis OX of the coordinate system associated with the spheroid. The relations between local coordinates (x, y, z) and global coordinates are expressed by the following formulas:

$$
x = Z_0 - Z, \qquad y = Y, \qquad z = X - X_0.
$$

Thus, the incident wave is TM polarized

$$
{}^{x}H_{\varphi} = \frac{k^2 D_x}{4\pi p} \frac{\sin\varphi}{\eta + \zeta_0} \mathrm{e}^{\mathrm{i}kp\eta - \mathrm{i}kZ_0 + \mathrm{i}\Phi - \mathrm{i}\Psi\cos\varphi},
$$

$$
{}^{x}E_{\varphi} = \frac{k^2 D_x}{4\pi p} \frac{\cos\varphi}{\eta + \zeta_0} \mathrm{e}^{\mathrm{i}kp\eta - \mathrm{i}kZ_0 + \mathrm{i}\Phi - \mathrm{i}\Psi\cos\varphi}.
$$

For the Fourier harmonics, this yields

$$
{}^{x}A_n^{\mathrm{inc}} = \Omega_x(-\mathrm{i})^{n-1} h_n(\eta, \tau), \qquad {}^{x}B_n^{\mathrm{inc}} = \Omega_x(-\mathrm{i})^{n-1} h_{n+2}(\eta, \tau), \tag{4.37}
$$

where

$$
h_n(\eta, \tau) = \frac{1}{\eta + \zeta_0} \exp\left(\frac{\mathrm{i}\chi}{2}\tau\eta + \mathrm{i}\Phi\right) J_{n-1}(\Psi),
$$

$$
\Omega_x = \frac{k^2 D_x}{4p} \mathrm{e}^{-\mathrm{i}kZ_0}.
$$

For the dipole with its momentum oriented along the axis OY, one can similarly get

$$
{}^{y}A_n^{\mathrm{inc}} = \Omega_y(-\mathrm{i})^{n-1} h_n(\eta, \tau), \qquad {}^{y}B_n^{\mathrm{inc}} = \Omega_y(-\mathrm{i})^{n-1} h_{n+2}(\eta, \tau), \tag{4.38}
$$

where

$$
\Omega_y = \frac{k^2 D_y}{4p} \mathrm{e}^{-\mathrm{i}kZ_0},
$$

but the field is TE polarized in this case.

Functions h_n satisfy parabolic equations $L_{n-1}h_n = 0$ and therefore can be represented in the form of the integrals (4.17) and (4.18), which will be done in the next section.

4.4.2.2 Matching Dipole Fields with Integral Representations

Functions a_n^{inc} and b_n^{inc} in the representations (4.17) and (4.18) for the incident field should be determined by equating these representations to the approximate expressions for the incident fields. This leads to integral equations

$$
\frac{1}{\sqrt{\chi\tau}\sqrt{1-\eta^2}} \int_{-\infty}^{+\infty} \left(\frac{1-\eta}{1+\eta}\right)^{\mathrm{i}t} M_{\mathrm{i}t, \frac{n\pm1}{2}}(-\mathrm{i}\chi\tau) w(t)\, \mathrm{d}t = \ldots, \tag{4.39}
$$

where $w(t)$ is the unknown function and the right-hand sides are functions f_n, g_n and h_n depending on the considered case of the dipole momentum orientation. The further derivations are based on the inversion of the integral operator in (4.39). However, in order to get explicit expressions for the solutions in terms of Whittaker functions, we perform the same trick which we used before in the case of similar integral equations. We exploit the fact that the solutions $w(t)$ do not depend on parameter τ presented in (4.39) and perform the following simplifications. We multiply the left-hand side and the right-hand side of the integral equations by τ^{δ} and let $\tau \to 0$. The power δ is chosen such that the limits are finite and not equal to zero.

The integral equations with the right-hand side expressed by functions h_n appeared already in the previous chapter, which enables us to use the solution found there. For the sake of consistency, it is represented as

$$w_n^h = \frac{(-1)^n}{\pi[(n-1)!]^2\sqrt{\chi\gamma_0}} \mathrm{e}^{-\mathrm{i}\alpha\eta_0/2} \left(\frac{\zeta_0 - 1}{\zeta_0 + 1}\right)^{\mathrm{i}t} \times$$
$$\times \Gamma\left(\tfrac{n}{2} + \mathrm{i}t\right) \Gamma\left(\tfrac{n}{2} - \mathrm{i}t\right) M_{\mathrm{i}t, \frac{n-1}{2}}(\mathrm{i}\Upsilon_0), \qquad (4.40)$$

where

$$\Upsilon_0 = \frac{\chi\gamma_0^2}{\zeta_0^2 - 1}.$$

For the dipole on the axis OZ, that is in the case of $\gamma_0 = 0$, by eliminating the ambiguity, it is easy to get that

$$w_1^h = -\frac{\mathrm{e}^{\mathrm{i}\pi/4}}{\sqrt{\zeta_0^2 - 1}} \frac{1}{\cosh(\pi t)} \left(\frac{\zeta_0 - 1}{\zeta_0 + 1}\right)^{\mathrm{i}t},$$

while all other $w_n^h = 0$, $n = 2, 3 \ldots$.

Amplitudes a_n^{inc} and b_n^{inc} are expressed via w^h in agreement with formulas (4.37) and (4.38)

$${}^{x}a_n^{\mathrm{inc}} = \Omega_x(-\mathrm{i})^{n-1} w_n,$$

$${}^{x}b_n^{\mathrm{inc}} = \Omega_x(-\mathrm{i})^{n-1} w_{n+2}.$$

And we remind that the electromagnetic field is TM polarized.

Consider the integral equation with the right-hand side given by function f_n which is defined by (4.36). In the case of $\tau \to 0$, we get

$$f_n \sim \frac{\gamma_0^n \chi^{n-1}}{2^{n-1}(n-1)!} \exp\left(\frac{\mathrm{i}\chi}{2} \frac{\gamma_0^2}{\eta + \zeta_0}\right) \frac{(1-\eta^2)^{\frac{n-1}{2}}}{(\eta + \zeta_0)^{n+1}} \tau^{\frac{n-1}{2}}.$$

Taking the limit as $\tau \to 0$, in view of the asymptotic representation of Bessel functions of small argument, we get the simplified integral equation in the form

$$\frac{1}{\sqrt{1-\eta^2}} \int_{-1}^{+1} \left(\frac{1-\eta}{1+\eta}\right)^{\mathrm{i}t} w_n^f(t)\mathrm{d}t = C_n \exp\left(\frac{\mathrm{i}\chi}{2} \frac{\gamma_0^2}{\eta+\zeta_0}\right) \frac{(1-\eta^2)^{\frac{n-1}{2}}}{(\eta+\zeta_0)^{n+1}}, \tag{4.41}$$

where

$$C_n = \frac{\mathrm{i}^{\frac{n}{2}} \chi^{\frac{n}{2}-1} \gamma_0^n}{2^{n-1}(n-1)!}.$$

Inverting the integral operator of (4.41) yields the integral

$$w_n^f = \frac{C_n}{\pi} \int_{-1}^{1} \left(\frac{1+\eta}{1-\eta}\right)^{\mathrm{i}t} \exp\left(\frac{\mathrm{i}\chi}{2} \frac{\gamma_0^2}{\eta+\zeta_0}\right) \frac{(1-\eta^2)^{\frac{n}{2}-1}}{(\eta+\zeta_0)^{n+1}}\, \mathrm{d}\eta. \tag{4.42}$$

Further derivations allow this integral to be expressed in terms of Whittaker functions. First, introduce new integration variable s, defined by the formulas

$$\eta = \frac{1-s}{1+s}, \qquad s = \frac{1-\eta}{1+\eta}, \qquad d\eta = -\frac{2}{(1+s)^2}\, ds.$$

Note also the

$$\eta + \zeta_0 = \frac{1-s}{1+s} + \zeta_0 = (\zeta_0 + 1)\frac{1+\Theta s}{1+s},$$

where

$$\Theta = \frac{\zeta_0 - 1}{\zeta_0 + 1} < 1.$$

Then the integral in (4.42) can be rewritten as

$$w_n^f = \frac{C_n}{\pi} \frac{2^{n-1}}{(\zeta_0+1)^{n+1}} \int_0^{+\infty} s^{\frac{n}{2}-\mathrm{i}t-1} \mathrm{e}^{\kappa(s)} \left(\frac{1+s}{1+\Theta s}\right)^{n+1} \frac{\mathrm{d}s}{(1+s)^n},$$

where

$$\kappa(s) = \frac{\mathrm{i}\chi\gamma_0^2}{2(\zeta_0+1)} \frac{1+s}{1+\Theta s}.$$

Now rewrite $\kappa(s)$ in the form

$$\kappa(s) = \frac{\mathrm{i}\chi\gamma_0^2}{2(\zeta_0-1)} - \frac{\mathrm{i}\alpha}{1+\Theta s}$$

and change the integration variable to $\sigma = \Theta s$. This yields

$$w_n^f = \frac{C_n}{\pi}\frac{2^{n-1}}{(\zeta_0+1)^{n+1}}\exp\left(\frac{i\chi\gamma_0^2}{2(\zeta_0-1)}\right)\Theta^{\mathrm{i}t-\frac{n+1}{2}}\times$$
$$\times\int_0^{+\infty}\sigma^{-\mathrm{i}t}\exp\left(-\frac{\mathrm{i}\alpha}{1+\sigma}\right)\left[\sqrt{\Theta}\sigma^{\frac{n}{2}-1}+\frac{\sigma^{\frac{n}{2}}}{\sqrt{\Theta}}\right]\frac{\mathrm{d}\sigma}{(1+\sigma)^{n+1}}.$$

Finally, use the integration variable ξ such that

$$\frac{1}{1+\sigma}=\frac{\xi+1}{2},$$

that is

$$\sigma=\frac{1-\xi}{1+\xi},\qquad d\sigma=-\frac{2}{(1+\xi)^2}\,d\xi.$$

Then

$$w_n^f=\frac{C_n}{2\pi}\frac{1}{(\zeta_0+1)^{n+1}}\exp\left(\frac{\mathrm{i}\chi\gamma_0^2}{2(\zeta_0-1)}-\frac{\mathrm{i}\Upsilon_0}{2}\right)\Theta^{\mathrm{i}t-\frac{n+1}{2}}\times$$
$$\times\left[\Theta^{\frac{1}{2}}\int_{-1}^{+1}(1+\xi)^{\mathrm{i}t+\frac{n}{2}}(1-\xi)^{-\mathrm{i}t+\frac{n}{2}-1}\exp\left(-\frac{\mathrm{i}\Upsilon_0\xi}{2}\right)\mathrm{d}\xi+\right.$$
$$\left.+\Theta^{-\frac{1}{2}}\int_{-1}^{+1}(1+\xi)^{\mathrm{i}t+\frac{n}{2}-1}(1-\xi)^{-\mathrm{i}t+\frac{n}{2}}\exp\left(-\frac{\mathrm{i}\Upsilon_0\xi}{2}\right)\mathrm{d}\xi\right]. \tag{4.43}$$

By comparing the integrals in (4.43) and in the integral representation (A.9) for the Whittaker function M, we finally get

$$w_n^f=\frac{\mathrm{e}^{-\mathrm{i}\pi/4}(-1)^{n-1}}{\pi(n-1)!n!\chi\gamma_0}\mathrm{e}^{\mathrm{i}\Upsilon_0\zeta_0/2}\Gamma\left(\tfrac{n}{2}+\mathrm{i}t\right)\Gamma\left(\tfrac{n}{2}-\mathrm{i}t\right)\times$$
$$\times\left(\frac{\zeta_0-1}{\zeta_0+1}\right)^{\mathrm{i}t}\left[\sqrt{\frac{\zeta_0-1}{\zeta_0+1}}\left(\frac{n}{2}+\mathrm{i}t\right)M_{\mathrm{i}t+\frac{1}{2},\frac{n}{2}}(\mathrm{i}\Upsilon_0)\right.$$
$$\left.+\sqrt{\frac{\zeta_0+1}{\zeta_0-1}}\left(\frac{n}{2}-\mathrm{i}t\right)M_{\mathrm{i}t-\frac{1}{2},\frac{n}{2}}(\mathrm{i}\Upsilon_0)\right].$$

Using relations for the Whittaker functions, this expression can be also rewritten as

$$w_n^f = \frac{(-1)^n}{\pi n!(n-1)!}\frac{e^{i\Upsilon_0\zeta_0/2}}{\zeta_0^2-1}\Gamma\left(\tfrac{n}{2}+it\right)\Gamma\left(\tfrac{n}{2}-it\right)\times$$
$$\times\left(\frac{\zeta_0-1}{\zeta_0+1}\right)^{it}\left[\frac{n^2+4t^2}{2n(n+1)}M_{it,\frac{n+1}{2}}(i\Upsilon_0)+(n\zeta_0-2it)M_{it,\frac{n-1}{2}}(i\Upsilon_0)\right]. \tag{4.44}$$

It is worth noting that in the case of $\gamma_0 = 0$, all the functions w_n^f nullify.

Consider now the integral equation with the right-hand side g_n. With the use of recurrence relations for Bessel functions [1]

$$J_{n-1}(z) + J_{n+1}(z) = \frac{2n}{z}J_n(z),$$

it is easy to find out that

$$g_n = f_{n+2} - \frac{2i(n+1)}{\chi\gamma_0}h_{n+2}.$$

Thus, it is possible to use the solutions (4.40) and (4.44) and obtain

$$w_n^g = w_{n+2}^f - \frac{2i(n+1)}{\chi\gamma_0}w_{n+2}^h. \tag{4.45}$$

In the case of $\gamma_0 = 0$, one needs to exclude the ambiguity in the second terms in (4.45). This results in

$$w_0^g = \frac{2}{\sqrt{\chi}(\zeta_0^2-1)}\frac{t}{\sinh(\pi t)}\left(\frac{\zeta_0-1}{\zeta_0+1}\right)^{it}$$

and $w_n^g = 0$ for $n = 1, 2, \ldots$

One can check that when the point of the dipole tends to infinity, the formulas of this section are reduced to the case of the incident plane wave.

4.5 Backward Wave

4.5.1 Representation of the Field Near the Trailing Tip

In the vicinity of the trailing tip of the spheroid, the surface can be approximated by a paraboloid. This enables us to approximate the scattered field by the field of diffraction by the perfectly conducting paraboloid whose surface is specified in paraboloidal coordinates (v, w), which are introduced in (4.4), by the equation $v = \chi$.

For further analysis, it is not important whether we consider the case of TE polarization with the representation (4.14) or the case of TM polarization with the representation (4.28), and we are able to deal with both cases simultaneously.

According to formulas (6.20) and (6.21) from [17], Fourier components of the fields can be written as

$$E_n = (vw)^{\frac{n-1}{2}} \left\{ Q^*_{n-1} - \frac{n}{2}\frac{\partial Q_{n-1}}{\partial w} + \frac{n}{2}\frac{\partial Q_{n-1}}{\partial v} - n\frac{\partial^2 P_{n-1}}{\partial v \partial w} + \frac{n}{4}P_{n-1} \right\}, \tag{4.46}$$

$$\mathrm{i}H_n = (uv)^{\frac{n-1}{2}} \left\{ P^*_{n-1} - \frac{n}{2}\frac{\partial P_{n-1}}{\partial w} + \frac{n}{2}\frac{\partial P_{n-1}}{\partial v} + n\frac{\partial^2 Q_{n-1}}{\partial v \partial w} - \frac{n}{4}Q_{n-1} \right\}, \tag{4.47}$$

where

$$P^*_{n-1} = w\frac{\partial^2 P_{n-1}}{\partial w^2} + n\frac{\partial P_{n-1}}{\partial w} + \frac{w}{4}P_{n-1} = -v\frac{\partial^2 P_{n-1}}{\partial v^2} - n\frac{\partial P_{n-1}}{\partial v} - \frac{v}{4}P_{n-1},$$

$$Q^*_{n-1} = w\frac{\partial^2 Q_{n-1}}{\partial w^2} + n\frac{\partial Q_{n-1}}{\partial w} + \frac{w}{4}Q_{n-1} = -v\frac{\partial^2 Q_{n-1}}{\partial v^2} - n\frac{\partial Q_{n-1}}{\partial v} - \frac{v}{4}Q_{n-1},$$

and P_{n-1} and Q_{n-1} satisfy the equations

$$\left[L_v + L_w + (n-1)\left(\frac{\partial}{\partial w} + \frac{\partial}{\partial v}\right)\right]\begin{pmatrix} P_{n-1} \\ Q_{n-1} \end{pmatrix} = 0, \tag{4.48}$$

where

$$L_v = v\frac{\partial^2}{\partial v^2} + \frac{\partial}{\partial v} + \frac{v}{4}, \qquad L_w = w\frac{\partial^2}{\partial w^2} + \frac{\partial}{\partial w} + \frac{w}{4}.$$

We choose the solutions of (4.48) in the form

$$P_{n-1} = \lambda(vw)^{-n/2} M_{\mathrm{i}t,\frac{n-1}{2}}(\mathrm{i}w) W_{\mathrm{i}t,\frac{n-1}{2}}(-\mathrm{i}v), \tag{4.49}$$

$$Q_{n-1} = \nu(vw)^{-n/2} M_{\mathrm{i}t,\frac{n-1}{2}}(\mathrm{i}w) W_{\mathrm{i}t,\frac{n-1}{2}}(-\mathrm{i}v). \tag{4.50}$$

Here, the regular Whittaker functions M are taken to specify the dependence on w coordinate, because the electromagnetic field should be regular at $w = 0$, and

Whittaker functions W are taken to specify the dependence of the field on v coordinate because of the radiation conditions.

Consider first the case $n > 0$. Using the properties of the Whittaker functions, one can check that $P_{s-1}^* = tP_{s-1}$ and $Q_{s-1}^* = tQ_{s-1}$. Substituting expressions (4.49), (4.50) in (4.46), we get

$$\begin{aligned} E_n = &\frac{1}{n\sqrt{vw}}\left(t^2+\frac{n^2}{4}\right)\Big[\lambda M_{\mathrm{i}t,\frac{n-1}{2}}(\mathrm{i}w) \\ &\qquad +\mathrm{i}\frac{\lambda t-\nu\frac{n}{2}}{n(n+1)}M_{\mathrm{i}t,\frac{n+1}{2}}(\mathrm{i}w)\Big]W_{\mathrm{i}t,\frac{n-1}{2}}(-\mathrm{i}v) \\ &-\frac{1}{n\sqrt{vw}}\left(t+\frac{in}{2}\right)\Big[\left(\lambda t-\nu\frac{n}{2}\right)M_{\mathrm{i}t,\frac{n-1}{2}}(\mathrm{i}w) \\ &\qquad +\frac{\mathrm{i}\lambda}{n(n+1)}\left(t^2+\frac{n^2}{4}\right)M_{\mathrm{i}t,\frac{n+1}{2}}(\mathrm{i}w)\Big]W_{\mathrm{i}t,\frac{n+1}{2}}(-\mathrm{i}v). \end{aligned} \tag{4.51}$$

The expression for $\mathrm{i}H_n$ differs by replacing ν with λ and λ with $-\nu$. The amplitudes λ and ν in the above formula are not arbitrary, but are related to each other by the requirement that the boundary conditions are satisfied. Particularly from the condition that $E_n = 0$ at $v = \chi$, we get the relations

$$\nu = \frac{\lambda}{n}\left[2t-\left(2t+in\right)\frac{W_{\mathrm{i}t,\frac{n+1}{2}}(-\mathrm{i}\chi)}{W_{\mathrm{i}t,\frac{n-1}{2}}(-\mathrm{i}\chi)}\right] \tag{4.52}$$

and

$$\nu = \frac{\lambda}{n}\left[2t-\left(2t-in\right)\frac{W_{\mathrm{i}t,\frac{n-1}{2}}(-\mathrm{i}\chi)}{W_{\mathrm{i}t,\frac{n+1}{2}}(-\mathrm{i}\chi)}\right]. \tag{4.53}$$

Requiring that there exists a nontrivial solution of system (4.52) and (4.53), we come to the dispersion equation for t:

$$(2t+in)\,W_{\mathrm{i}t,\frac{n+1}{2}}^2(-\mathrm{i}\chi) = (2t-in)\,W_{\mathrm{i}t,\frac{n-1}{2}}^2(-\mathrm{i}\chi). \tag{4.54}$$

One can check that the requirement of the second boundary condition to be satisfied results in the same relations (4.52) and (4.53).

Consider now the case of $n = 0$. In this case, formulas (4.46) and (4.47) simplify to

$$E_0 = \frac{\nu t}{\sqrt{vw}}M_{\mathrm{i}t,1/2}(\mathrm{i}w)W_{\mathrm{i}t,1/2}(-\mathrm{i}v), \tag{4.55}$$

$$\mathrm{i}H_0 = \frac{\lambda t}{\sqrt{vw}}M_{\mathrm{i}t,1/2}(\mathrm{i}w)W_{\mathrm{i}t,1/2}(-\mathrm{i}v).$$

The requirements for the boundary conditions to be satisfied result in the two types of solutions. In the first case, the amplitude λ is equal to zero and t satisfies the dispersion equation

$$W_{\mathrm{i}t,1/2}(-\mathrm{i}\chi) = 0. \tag{4.56}$$

In the second case, $\nu = 0$ and t satisfies the equation

$$W'_{\mathrm{i}t,1/2}(-\mathrm{i}\chi) = 0. \tag{4.57}$$

The dispersion equations (4.54), (4.56) and (4.57) define a discrete set of complex values t_j for every n. The component E_φ is the sum of such solutions. That is, to define the electromagnetic field one needs to specify all the amplitudes $\lambda_{n,j}$, $\nu_{n,j}$.

4.5.2 Determining the Amplitudes $\lambda_{n,j}$ and $\nu_{n,j}$

The representations of Sect. 4.3 are valid in the middle part of the boundary layer. When approaching the trailing tip, i.e. when $\eta \to 1$ these representations are not valid. On the contrary, the representation (4.51) is valid when the point of observation is close to the tip and becomes wrong when w increases too much. However, there exists a matching region, where both representations are still valid. This region is characterized by the conditions $w \gg 1$, $1 - \eta \ll 1$. Relations (4.5) show that such a domain exists. We match the representations in this domain and by this determine the amplitudes $\lambda_{n,j}$ and $\nu_{n,j}$.

It is worth noting that the multiplier $((1-\eta)/(1+\eta))^{\mathrm{i}t}$ in the integrals (4.15) and (4.16) can be rewritten as

$$\exp(\mathrm{i}Xt), \qquad X = \ln\left(\frac{1-\eta}{1+\eta}\right),$$

which shows that for $\eta \to 1$, when the factor X becomes large negative, the path of integration can be closed by adding an arc of large radius in the lower half-plane of t. Thus, by residue theorem the integrals are converted to the sums of residues in the poles. These poles are the solutions of the dispersion equation $Z_n = 0$, and the same equation determines the parameters $t_{n,j}$ in (4.51).

In (4.51), we have the Whittaker function M of the large argument $\mathrm{i}w$ and can replace it with the asymptotic representation

$$M_{\mathrm{i}t,\mu}(\mathrm{i}w) \sim \frac{\Gamma(2\mu+1)\mathrm{e}^{\pi t/2}}{\Gamma\left(\mu+\frac{1}{2}-\mathrm{i}t\right)} w^{-\mathrm{i}t}\mathrm{e}^{\mathrm{i}w/2} + \frac{\Gamma(2\mu+1)\mathrm{e}^{\pi t/2}}{\Gamma\left(\mu+\frac{1}{2}+\mathrm{i}t\right)} \mathrm{e}^{\mathrm{i}\pi\left(\mu+\frac{1}{2}\right)} w^{\mathrm{i}t}\mathrm{e}^{-\mathrm{i}w/2}. \tag{4.58}$$

This representation follows from formulas 9.233.2 and 9.227 from [18].

By substituting from (4.58) to (4.51) and grouping terms, we get

$$\begin{aligned} E_n \sim \frac{\Gamma(n)}{2\sqrt{vw}} \mathrm{e}^{\pi t/2} \Bigg\{ & \frac{w^{-\mathrm{i}t}\mathrm{e}^{\mathrm{i}w/2}}{\Gamma\left(\frac{n}{2}-\mathrm{i}t\right)} \left[\left(\frac{n}{2}+\mathrm{i}t\right)(\lambda - \mathrm{i}\nu) W_{\mathrm{i}t,\frac{n-1}{2}}(-\mathrm{i}v) \right. \\ & \left. + \left(\frac{n}{2}-\mathrm{i}t\right)(\lambda + \mathrm{i}\nu) W_{\mathrm{i}t,\frac{n+1}{2}}(-\mathrm{i}v) \right] \\ & + \mathrm{e}^{\mathrm{i}\pi n/2} \frac{w^{\mathrm{i}t}\mathrm{e}^{-\mathrm{i}w/2}}{\Gamma\left(\frac{n}{2}+\mathrm{i}t\right)} \left[\left(\frac{n}{2}-\mathrm{i}t\right)(\lambda + \mathrm{i}\nu) W_{\mathrm{i}t,\frac{n-1}{2}}(-\mathrm{i}v) \right. \\ & \left. - \left(\frac{n}{2}-\mathrm{i}t\right)(\lambda - \mathrm{i}\nu) W_{\mathrm{i}t,\frac{n+1}{2}}(-\mathrm{i}v) \right] \Bigg\}. \end{aligned} \tag{4.59}$$

In order to find the asymptotic expansions of (4.15) and (4.16) for $\eta \to 1$, we note that in view of (4.5) the following approximation holds:

$$\frac{\mathrm{e}^{\mathrm{i}kp\eta}}{\sqrt{1-\eta^2}} \left(\frac{1-\eta}{1+\eta}\right)^{\mathrm{i}t} \sim \frac{\sqrt{kb}}{\sqrt{w}} \mathrm{e}^{\mathrm{i}kp} w^{\mathrm{i}t} \mathrm{e}^{-\mathrm{i}w/2} (4kp)^{-\mathrm{i}t}.$$

This shows that the second term in (4.59) corresponds to the forward wave. By matching the amplitudes, we get the relations

$$\begin{aligned} \frac{\Gamma(n)\mathrm{e}^{\pi t_{n,j}/2}\mathrm{e}^{\mathrm{i}\pi n/2}}{2\Gamma\left(\frac{n}{2}+\mathrm{i}t_{n,j}\right)} & \left(\frac{n}{2}-\mathrm{i}t_{n,j}\right)(\lambda_{n,j}+\mathrm{i}\nu_{n,j}) \\ & = 2\pi\mathrm{i} \operatorname*{Res}_{t=t_{n,j}} \left(a_n^{\mathrm{forward}}(t)\sqrt{kp}\mathrm{e}^{\mathrm{i}kp}(4kp)^{-\mathrm{i}t}\right), \end{aligned}$$

$$\begin{aligned} \frac{\Gamma(n)\mathrm{e}^{\pi t_{n,j}/2}\mathrm{e}^{\mathrm{i}\pi n/2}}{2\Gamma\left(\frac{n}{2}+\mathrm{i}t_{n,j}\right)} & \left(\frac{n}{2}-\mathrm{i}t_{n,j}\right)(-\lambda_{n,j}+\mathrm{i}\nu_{n,j}) \\ & = 2\pi\mathrm{i} \operatorname*{Res}_{t=t_{n,j}} \left(b_n^{\mathrm{forward}}(t)\sqrt{kp}\mathrm{e}^{\mathrm{i}kp}(4kp)^{-\mathrm{i}t}\right), \end{aligned}$$

which determines the amplitudes $\lambda_{n,j}$ and $\nu_{n,j}$ in (4.51):

$$\begin{aligned} \lambda_{n,j} &= 2\pi\mathrm{i} \operatorname*{Res}_{t=t_{n,j}} \left(a_n^{\mathrm{forward}}(t) - b_n^{\mathrm{forward}}(t)\right)\varphi_n(t), \\ \nu_{n,j} &= 2\pi \operatorname*{Res}_{t=t_{n,j}} \left(a_n^{\mathrm{forward}}(t) + b_n^{\mathrm{forward}}(t)\right)\varphi_n(t), \end{aligned}$$

where

$$\varphi_n(t) = \sqrt{kp}\mathrm{e}^{\mathrm{i}kp}(4kp)^{-\mathrm{i}t} \frac{\mathrm{e}^{-\pi t/2}\mathrm{e}^{-\mathrm{i}\pi n/2}}{\Gamma(n)} \frac{\Gamma\left(\frac{n}{2}+\mathrm{i}t\right)}{\frac{n}{2}-\mathrm{i}t}.$$

Consider now the case of $n = 0$. Computing the asymptotics of (4.55) for $w \to +\infty$, we find

$$E_{\varphi}^{(0)} \sim \frac{\nu t e^{\pi t/2}}{\sqrt{vw}} \left\{ \frac{w^{-\mathrm{i}t} e^{\mathrm{i}w/2}}{\Gamma(1-\mathrm{i}t)} - \frac{w^{\mathrm{i}t} e^{-\mathrm{i}w/2}}{\Gamma(1+\mathrm{i}t)} \right\} W_{\mathrm{i}t,1/2}(-\mathrm{i}v). \tag{4.60}$$

By matching the forward wave with the second term in (4.60), we get

$$-t_{0,j}\nu_{0,j} \frac{e^{\pi t_{0,j}/2}}{\Gamma(1+\mathrm{i}t_{0,j})} = 2\pi\mathrm{i} \operatorname*{Res}_{t=t_{0,j}} b_0^{\text{forward}}(t)\sqrt{kb} e^{\mathrm{i}kp}(4kp)^{-\mathrm{i}t}, \tag{4.61}$$

which defines the amplitudes $\nu_{0,j}$.

The first terms in (4.59) and (4.60) correspond to the backward wave, which we study in the next section.

4.5.3 Representation for the Backward Wave

Consider the backward wave. This wave propagates in the negative direction of z, and we represent it in the middle part of the surface in the form

$$E_n^{\text{back}} = e^{-\mathrm{i}kp\eta} \mathcal{E}_n^{\text{back}}(\eta, \tau), \qquad H_n^{\text{back}} = e^{-\mathrm{i}kp\eta} \mathcal{H}_n^{\text{back}}(\eta, \tau), \tag{4.62}$$

where $\mathbf{E}_n^{\text{back}}$ and $\mathbf{H}_n^{\text{back}}$ satisfy the system of parabolic equations. By using the representation (4.12), we separate these equations for the unknowns A_n^{back} and B_n^{back} and get

$$L_{n-1}^{\text{back}} A_n^{\text{back}} = 0, \qquad L_{n+1}^{\text{back}} B_n^{\text{back}} = 0, \tag{4.63}$$

where the parabolic operator L_s^{back} differs from (4.10) by η being replaced with $-\eta$.

We solve equations (4.63) by variable separation method, however, instead of the integral with respect to the parameter t, we use summation in agreement with the representation of the field near the trailing tip.

So, in the case of TE polarization we have formulas (4.14), (4.62) and (4.12), where $A_0^{\text{back}} = 0$,

$$A_n^{\text{back}} = \frac{(\chi\tau)^{-1/2}}{\sqrt{1-\eta^2}} \sum_j \left(\frac{1+\eta}{1-\eta}\right)^{\mathrm{i}t_{n,j}} a_{n,j}^{\text{back}} W_{\mathrm{i}t_{n,j},\frac{n-1}{2}}(-\mathrm{i}\chi\tau), \tag{4.64}$$

$$B_n^{\text{back}} = \frac{(\chi\tau)^{-1/2}}{\sqrt{1-\eta^2}} \sum_j \left(\frac{1+\eta}{1-\eta}\right)^{\mathrm{i}t_{n,j}} b_{n,j}^{\text{back}} W_{\mathrm{i}t_{n,j},\frac{n+1}{2}}(-\mathrm{i}\chi\tau). \tag{4.65}$$

The amplitudes $a_{n,t}^{\text{back}}$ and $b_{n,t}^{\text{back}}$ should be taken such that the boundary conditions are satisfied. For the backward wave, these conditions are homogeneous. Therefore, for the Fourier harmonics with $n = 0$ the dispersion equation should hold

$$W_{\mathrm{i}t_{0,j},\frac{1}{2}}(-\mathrm{i}\chi) = 0 \tag{4.66}$$

with respect to the parameter $t_{0,j}$, and for $n > 0$ the amplitudes $a_{n,j}^{\text{back}}$ and $b_{n,j}^{\text{back}}$ should satisfy the relation

$$a_{n,j}^{\text{back}} W_{\mathrm{i}t_{n,j},\frac{n-1}{2}}(-\mathrm{i}\chi) + b_{n,j}^{\text{back}} W_{\mathrm{i}t_{n,j},\frac{n+1}{2}}(-\mathrm{i}\chi) = 0, \tag{4.67}$$

while $t_{n,j}$ are the solutions of the equation

$$\begin{aligned} W_{\mathrm{i}t_{n,j},\frac{n-1}{2}}(-\mathrm{i}\chi) W'_{\mathrm{i}t_{n,j},\frac{n+1}{2}}(-\mathrm{i}\chi) \\ + W'_{\mathrm{i}t_{n,j},\frac{n-1}{2}}(-\mathrm{i}\chi) W_{\mathrm{i}t_{n,j},\frac{n+1}{2}}(-\mathrm{i}\chi) = 0. \end{aligned} \tag{4.68}$$

Similarly, in the case of TM polarization we have Fourier series (4.28) with the components defined by expressions (4.62), (4.12), (4.64) and (4.65). For $n = 0$, the dispersion equation (4.66) is replaced with

$$W'_{\mathrm{i}t_{0,j},\frac{1}{2}}(-\mathrm{i}\chi) = 0 \tag{4.69}$$

and formulas (4.67) and (4.68) for $n > 0$ remain valid.

The amplitudes $a_{n,j}^{\text{back}}$ and $b_{n,j}^{\text{back}}$ of the backward wave can be found from the asymptotics (4.59). Taking into account that

$$\frac{\mathrm{e}^{-ikp\eta}}{\sqrt{1-\eta^2}} \left(\frac{1+\eta}{1-\eta}\right)^{\mathrm{i}t} \sim \frac{\sqrt{kb}}{\sqrt{w}} \mathrm{e}^{-ikp} w^{-\mathrm{i}t} \mathrm{e}^{\mathrm{i}w/2} (4kp)^{\mathrm{i}t}$$

and performing similar derivations as in Sect. 4.5.2, we get

$$\begin{aligned} a_{n,j}^{\text{back}} &= -2\pi \mathrm{i} \mathrm{e}^{2ikp - \mathrm{i}\pi n/2} \operatorname*{Res}_{t=t_{n,j}} \left(b_n^{\text{forward}}(t) \frac{\Gamma\left(\frac{n}{2} + 1 + \mathrm{i}t\right)}{\Gamma\left(\frac{n}{2} + 1 - \mathrm{i}t\right)} (4kp)^{-2\mathrm{i}t} \right), \\ b_{n,j}^{\text{back}} &= 2\pi \mathrm{i} \mathrm{e}^{2ikp - \mathrm{i}\pi n/2} \operatorname*{Res}_{t=t_{n,j}} \left(a_n^{\text{forward}}(t) \frac{\Gamma\left(\frac{n}{2} + \mathrm{i}t\right)}{\Gamma\left(\frac{n}{2} - \mathrm{i}t\right)} (4kp)^{-2\mathrm{i}t} \right). \end{aligned} \tag{4.70}$$

In the case of $n = 0$, we match the first term in (4.60) with the backward wave, which requires that

$$t_{0,j} \nu_{0,j} \frac{\mathrm{e}^{\pi t_{0,j}/2}}{\Gamma(1 - \mathrm{i}t_{0,j})} = b_{0,j}^{\text{back}} \sqrt{kb} \mathrm{e}^{-ikp} (4kp)^{\mathrm{i}t_{0,j}},$$

and together with (4.61) allows the amplitudes of the backward wave to be found as

$$b_{0,j}^{\text{back}} = -2\pi \mathrm{i} \operatorname*{Res}_{t=t_{0,j}} \left(b^{\text{forward}}(t) \frac{\Gamma(1+\mathrm{i}t)}{\Gamma(1-\mathrm{i}t)} \mathrm{e}^{2\mathrm{i}kp} (4kp)^{-2\mathrm{i}t} \right). \tag{4.71}$$

Note that due to the factor $(4kp)^{-2\mathrm{i}t_{n,j}}$, the series (4.64) and (4.65) converge for any $\eta \in (-1, 1)$.

We have found the amplitudes of the backward wave by matching the expressions for the E_φ components only. However, it can be checked that all the other components of the electromagnetic fields also match if the amplitudes λ and ν are taken according to expressions (4.52) and (4.53) and the amplitudes a_n^{back} and b_n^{back} of the backward wave are taken according to expressions (4.67).

4.6 Induced Currents

One can calculate the induced current $J = H_\varphi(\eta, 1)$, where $H_\varphi(\eta, 1)$ is the angular component of the total magnetic field on the surface. The forward component of the current is formed by the incident and by the forward waves,

$$J_n^{\text{forward}} = \frac{\mathrm{e}^{\mathrm{i}kp\eta}}{\sqrt{\chi}\sqrt{1-\eta^2}} \int_{-\infty}^{+\infty} \left(\frac{1-\eta}{1+\eta} \right)^{\mathrm{i}t} \left\{ a_n^{\text{inc}} M_{\mathrm{i}t, \frac{n-1}{2}}(-\mathrm{i}\chi) \right.$$
$$\left. - b_n^{\text{inc}} M_{\mathrm{i}t, \frac{n+1}{2}}(-\mathrm{i}\chi) + a_n^{\text{forward}} W_{\mathrm{i}t, \frac{n-1}{2}}(-\mathrm{i}\chi) - b_n^{\text{forward}} W_{\mathrm{i}t, \frac{n+1}{2}}(-\mathrm{i}\chi) \right\} \mathrm{d}t. \tag{4.72}$$

By substituting here expressions (4.22) and (4.23) and taking into account formula (4.21), it is a simple matter to get

$$J_n^{\text{forward}} = -\frac{2\,\mathrm{e}^{\mathrm{i}kp\eta}}{\sqrt{\chi}\sqrt{1-\eta^2}} \int_{-\infty}^{+\infty} \left(\frac{1-\eta}{1+\eta} \right)^{\mathrm{i}t} \left\{ a_n^{\text{inc}} \frac{(n-1)!}{\Gamma\left(\frac{n}{2} - \mathrm{i}t\right)} W_{\mathrm{i}t, \frac{n+1}{2}}(-\mathrm{i}\chi) \right.$$
$$\left. - b_n^{\text{inc}} \frac{(n+1)!}{\Gamma\left(\frac{n}{2} + 1 - \mathrm{i}t\right)} W_{\mathrm{i}t, \frac{n-1}{2}}(-\mathrm{i}\chi) \right\} \frac{\mathrm{d}t}{Z_n}. \tag{4.73}$$

For $n = 0$, the nonzero current appears only in the case of TM polarization

$$J_0^{\text{forward}} = -\frac{\mathrm{e}^{\mathrm{i}kp\eta}}{\sqrt{\chi}\sqrt{1-\eta^2}} \int_{-\infty}^{+\infty} \left(\frac{1-\eta}{1+\eta} \right)^{\mathrm{i}t} \times$$
$$\times \left\{ b_0^{\text{inc}} M_{\mathrm{i}t, \frac{1}{2}}(-\mathrm{i}\chi) + b_0^{\text{forward}} W_{\mathrm{i}t, \frac{1}{2}}(-\mathrm{i}\chi) \right\} \mathrm{d}t,$$

which by substituting from (4.29) yields

$$J_0^{\text{forward}} = \frac{\mathrm{e}^{ikp\eta}}{\sqrt{\chi}\sqrt{1-\eta^2}} \int_{-\infty}^{+\infty} \left(\frac{1-\eta}{1+\eta}\right)^{it} \frac{b_0^{\text{inc}}\, \mathrm{d}t}{\Gamma\left(1-\mathrm{i}t\right) W'_{\mathrm{i}t,\frac{1}{2}}(-\mathrm{i}\chi)}. \tag{4.74}$$

Formulas (4.73) and (4.74) represent the current of the forward wave. For the current of the backward wave, we get

$$J_n^{\text{back}} = \frac{\mathrm{e}^{-ikp\eta}}{\sqrt{\chi}\sqrt{1-\eta^2}} \sum_j \left(\left(\frac{1+\eta}{1-\eta}\right)^{\mathrm{i}t_{n,j}} \times \right.$$
$$\left. \times \left\{ a_{n,j}^{\text{back}} W_{\mathrm{i}t_{n,j},\frac{n-1}{2}}(-\mathrm{i}\chi) - b_{n,j}^{\text{back}} W_{\mathrm{i}t_{n,j},\frac{n+1}{2}}(-\mathrm{i}\chi) \right\} \right). \tag{4.75}$$

We can substitute here expressions (4.70) for the amplitudes $a_{n,j}^{\text{back}}$ and $b_{n,j}^{\text{back}}$ of backward waves. However, it is not easy to compute the field by this formula, because one needs to find the solutions $t_{n,j}$ of the dispersion equations (4.68) and (4.66) or (4.69). Moreover, there is no simple expression for the derivative of the Whittaker function W with respect to its first index, and such derivatives are present in the expression for dZ_n/dt contained in the denominator of (4.70). To this end, it is convenient to consider the sum in (4.75) as the sum of residues and replace it with the integral similar to (4.73).

For that, we need to represent the integrated function in such a form that it has no other poles except at $t = t_{n,j}$. First, we substitute expressions (4.70) and represent (4.75) as

$$J_n^{\text{back}} = -\frac{2\pi \mathrm{e}^{ikp(2-\eta)-\mathrm{i}\pi\frac{n-1}{2}}}{\sqrt{\chi}\sqrt{1-\eta^2}} \sum_j \operatorname*{Res}_{t=t_{n,j}} \left(\left(\frac{1+\eta}{1-\eta}\right)^{it} (4kp)^{-2\mathrm{i}t} \frac{\Gamma\left(\frac{n}{2}+\mathrm{i}t\right)}{\Gamma\left(\frac{n}{2}-\mathrm{i}t\right)} \times \right.$$
$$\left. \times \left\{ b_n^{\text{forward}}(t) \frac{\frac{n+2}{2}+\mathrm{i}t}{\frac{n+2}{2}-\mathrm{i}t} W_{\mathrm{i}t,\frac{n-1}{2}}(-\mathrm{i}\chi) + a_n^{\text{forward}}(t) W_{\mathrm{i}t,\frac{n+1}{2}}(-\mathrm{i}\chi) \right\} \right).$$

The expression in braces can be transformed to

$$\{\ldots\} = \frac{W_{\mathrm{i}t,\frac{n+1}{2}}(-\mathrm{i}\chi)}{W_{\mathrm{i}t,\frac{n-1}{2}}(-\mathrm{i}\chi)} \times$$
$$\times \left\{ a_n^{\text{forward}}(t) W_{\mathrm{i}t,\frac{n-1}{2}}(-\mathrm{i}\chi) + Q_n(t) b_n^{\text{forward}}(t) W_{\mathrm{i}t,\frac{n+1}{2}}(-\mathrm{i}\chi) \right\},$$

$$Q_n(t) = \frac{\frac{n}{2}+it}{\frac{n}{2}+it} \frac{W^2_{\mathrm{i}t,(n-1)/2}(-\mathrm{i}\chi)}{W^2_{\mathrm{i}t,(n+1)/2}(-\mathrm{i}\chi)} = \frac{nZ_n}{W^2_{\mathrm{i}t,(n+1)/2}(-\mathrm{i}\chi)} - 1.$$

Under the sign of the residue, one can take $Q_n(t) = -1$, which yields

$$\{\ldots\} = \frac{W_{\mathrm{i}t,\frac{n+1}{2}}(-\mathrm{i}\chi)}{W_{\mathrm{i}t,\frac{n-1}{2}}(-\mathrm{i}\chi)}\left\{a_n^{\mathrm{forward}}(t)W_{\mathrm{i}t,\frac{n-1}{2}}(-\mathrm{i}\chi) - b_n^{\mathrm{forward}}(t)W_{\mathrm{i}t,\frac{n+1}{2}}(-\mathrm{i}\chi)\right\}.$$

We need to substitute here expressions (4.22) and (4.23). However, if we add in the braces the function

$$a_n^{\mathrm{inc}}(t)M_{\mathrm{i}t,\frac{n-1}{2}}(-\mathrm{i}\chi) - b_n^{\mathrm{inc}}(t)M_{\mathrm{i}t,\frac{n+1}{2}}(-\mathrm{i}\chi)$$

which has no poles at $t = t_{n,j}$, we can use the result of such derivations performed when passing from (4.72) to (4.73). Thus, we get

$$\begin{aligned} J_n^{\mathrm{back}} = \frac{4\pi\mathrm{i}\,\mathrm{e}^{2\mathrm{i}kp-\mathrm{i}kp\eta-\mathrm{i}\pi n/2}}{\sqrt{\chi}\sqrt{1-\eta^2}} \sum_j \operatorname*{Res}_{t=t_{n,j}} \Bigg(\left(\frac{1+\eta}{1-\eta}\right)^{\mathrm{i}t} \frac{(4kp)^{-2\mathrm{i}t}}{Z_n(t)} \frac{\Gamma\left(\frac{n}{2}+\mathrm{i}t\right)}{\Gamma\left(\frac{n}{2}-\mathrm{i}t\right)} \times \\ \times \Bigg\{ a_n^{\mathrm{inc}}(t)\frac{(n-1)!}{\Gamma\left(\frac{n}{2}-\mathrm{i}t\right)} \frac{W^2_{\mathrm{i}t,(n+1)/2}(-\mathrm{i}\chi)}{W_{\mathrm{i}t,\frac{n-1}{2}}(-\mathrm{i}\chi)} \\ -b_n^{\mathrm{inc}}(t)\frac{(n+1)!}{\Gamma\left(\frac{n}{2}+1-\mathrm{i}t\right)} W_{\mathrm{i}t,\frac{n+1}{2}}(-\mathrm{i}\chi)\Bigg\}\Bigg). \end{aligned}$$

In this expression, the Whittaker function $W_{\mathrm{i}t,\frac{n-1}{2}}(-\mathrm{i}\chi)$ presented in the denominator adds the poles. Therefore, it is necessary to exclude it from the denominator. This can be done with the help of the relation (4.25) and suppressing the term that does not contain the denominator $Z_n(t)$. The resulting expression can be rewritten in the form of the integral

$$\begin{aligned} J_n^{\mathrm{back}} = \frac{2(-\mathrm{i})^n \mathrm{e}^{2\mathrm{i}kp-\mathrm{i}kp\eta}}{\sqrt{\chi}\sqrt{1-\eta^2}} \int_{\infty}^{+\infty} \left(\frac{1+\eta}{1-\eta}\right)^{\mathrm{i}t} \frac{(4kp)^{-2\mathrm{i}t}}{Z_n\Gamma\left(\frac{n}{2}-\mathrm{i}t\right)\Gamma\left(\frac{n}{2}+1-\mathrm{i}t\right)} \times \\ \times \left\{a_n^{\mathrm{inc}}(t)(n-1)!\Gamma\left(\frac{n}{2}+1+\mathrm{i}t\right) W_{\mathrm{i}t,\frac{n-1}{2}}(-\mathrm{i}\chi) \right. \\ \left. +b_n^{\mathrm{inc}}(t)(n+1)!\Gamma\left(\frac{n}{2}+\mathrm{i}t\right) W_{\mathrm{i}t,\frac{n+1}{2}}(-\mathrm{i}\chi)\right\} \mathrm{d}t. \end{aligned} \tag{4.76}$$

In the case of $n = 0$, requiring separate consideration, we get directly from (4.71) the desired representation. It can be at once rewritten in the form of the integral

$$J_0^{\mathrm{back}} = \frac{\mathrm{e}^{2\mathrm{i}kp-\mathrm{i}kp\eta}}{\sqrt{\chi}\sqrt{1-\eta^2}} \int_{\infty}^{+\infty} \left(\frac{1+\eta}{1-\eta}\right)^{\mathrm{i}t} (4kp)^{-2\mathrm{i}t} \frac{\Gamma(1+\mathrm{i}t)}{\Gamma(1-\mathrm{i}t)} b_0^{\mathrm{forward}}(t) W_{\mathrm{i}t,\frac{1}{2}}(-\mathrm{i}\chi)\mathrm{d}t.$$

We substitute here from (4.19), and express M' from (4.21). Suppressing the terms that have no poles in $\Im(t) < 0$, we finally get

$$J_0^{\text{back}} = -\frac{e^{2ikp - ikp\eta}}{\sqrt{\chi}\sqrt{1-\eta^2}} \int\limits_{\infty}^{+\infty} \left(\frac{1+\eta}{1-\eta}\right)^{it} (4kp)^{-2it} \frac{\Gamma(1+it)}{\Gamma^2(1-it)} \frac{b_0^{\text{inc}}(t)}{\dot{W}_{it,\frac{1}{2}}(-i\chi)} dt. \quad (4.77)$$

The integrals in formulas (4.76) and (4.77) exponentially converge and are easy to be computed.

4.7 Validation Tests

For the computations, we use the code developed in [21] for Coulomb wave functions F and H^+, [1]. For that reason, it is necessary to rewrite expressions (4.73) and (4.76) by representing the Whittaker functions in terms of Coulomb wave functions. The following formulas hold

$$M_{it,\ell}\left(i\alpha^2\right) = \frac{2\Gamma(2\ell+1)e^{\pi t/2}e^{i\pi(2\ell+1)/4}}{\sqrt{\Gamma\left(\ell+\frac{1}{2}+it\right)\Gamma\left(\ell+\frac{1}{2}-it\right)}} F_{\ell-\frac{1}{2}}\left(t, \frac{\alpha^2}{2}\right),$$

$$W_{it,\ell}(-i\chi) = e^{\pi t/2}e^{i\pi(2\ell-1)/4}\sqrt{\frac{\Gamma\left(\ell+\frac{1}{2}+it\right)}{\Gamma\left(\ell+\frac{1}{2}-it\right)}} H^+_{\ell-\frac{1}{2}}\left(-t, \frac{\chi}{2}\right).$$

This yields

$$Z_n = \frac{i^n}{2}e^{\pi t}\sqrt{\frac{\Gamma\left(\frac{n}{2}+it\right)\Gamma\left(\frac{n}{2}+1+it\right)}{\Gamma\left(\frac{n}{2}-it\right)\Gamma\left(\frac{n}{2}+1-it\right)}} \times$$
$$\times \left[H^+_{\frac{n}{2}}\left(-t, \tfrac{\chi}{2}\right)\dot{H}^+_{\frac{n}{2}-1}\left(-t, \tfrac{\chi}{2}\right) + \dot{H}^+_{\frac{n}{2}}\left(-t, \tfrac{\chi}{2}\right)H^+_{\frac{n}{2}-1}\left(-t, \tfrac{\chi}{2}\right)\right].$$

As the result, the leading order approximation for the current of the forward wave in the case of TE polarization takes the form

$$J_n^{\text{forward}} = \frac{8i^{n-1}e^{ikp\eta}}{\pi\sqrt{\chi}\sqrt{1-\eta^2}\alpha} \int\limits_{-\infty}^{+\infty} \left(\frac{1-\eta}{1+\eta}\right)^{it} Q_+(t)dt,$$

where

$$Q_\pm = \frac{F_{\frac{n}{2}-1}\left(t, \frac{\alpha^2}{2}\right)H^+_{\frac{n}{2}}\left(-t, \frac{\chi}{2}\right) \pm F_{\frac{n}{2}}\left(t, \frac{\alpha^2}{2}\right)H^+_{\frac{n}{2}-1}\left(-t, \frac{\chi}{2}\right)}{\dot{H}^+_{\frac{n}{2}-1}\left(-t, \frac{\chi}{2}\right)H^+_{\frac{n}{2}}\left(-t, \frac{\chi}{2}\right) + H^+_{\frac{n}{2}-1}\left(-t, \frac{\chi}{2}\right)\dot{H}^+_{\frac{n}{2}}\left(-t, \frac{\chi}{2}\right)}.$$

Here, dot over F or H_+ denotes the derivative of the function with respect to its second argument. The current of the backward wave is given by

$$J_n^{\text{back}} = \frac{8\,e^{2ikp-ikp\eta}}{\pi\sqrt{\chi}\sqrt{1-\eta^2}\alpha}\int_{-\infty}^{+\infty}\left(\frac{1+\eta}{1-\eta}\right)^{it}\tilde{Q}_+(t)\times$$

$$\times\,(4kp)^{-2it}\frac{\Gamma\left(\frac{n}{2}+it\right)}{\Gamma\left(\frac{n}{2}-it\right)}\sqrt{\frac{n+2it}{n-2it}}dt,$$

where

$$\tilde{Q}_\pm = \frac{F_{\frac{n}{2}-1}\left(t,\frac{\alpha^2}{2}\right)H^+_{\frac{n}{2}-1}\left(-t,\frac{\chi}{2}\right)\pm F_{\frac{n}{2}}\left(t,\frac{\alpha^2}{2}\right)H^+_{\frac{n}{2}}\left(-t,\frac{\chi}{2}\right)}{\dot{H}^+_{\frac{n}{2}-1}\left(-t,\frac{\chi}{2}\right)H^+_{\frac{n}{2}}\left(-t,\frac{\chi}{2}\right)+H^+_{\frac{n}{2}-1}\left(-t,\frac{\chi}{2}\right)\dot{H}^+_{\frac{n}{2}}\left(-t,\frac{\chi}{2}\right)}.$$

In the case of TM polarization, the expressions are similar:

$$J_0^{\text{forward}} = \frac{-4\,e^{ikp\eta}}{\pi\sqrt{\chi}\sqrt{1-\eta^2}\alpha}\int_{-\infty}^{+\infty}\left(\frac{1-\eta}{1+\eta}\right)^{it}\frac{F_0\left(t,\frac{\alpha^2}{2}\right)}{\dot{H}_0^+\left(-t,\frac{\chi}{2}\right)}dt, \tag{4.78}$$

$$J_n^{\text{forward}} = \frac{-8i^n e^{ikp\eta}}{\pi\sqrt{\chi}\sqrt{1-\eta^2}\alpha}\int_{-\infty}^{+\infty}\left(\frac{1-\eta}{1+\eta}\right)^{it}Q_-(t)dt, \tag{4.79}$$

$$J_0^{\text{back}} = \frac{4\,e^{2ikp-ikp\eta}}{\pi\sqrt{\chi}\sqrt{1-\eta^2}\alpha}\int_{-\infty}^{+\infty}\left(\frac{1+\eta}{1-\eta}\right)^{it}\times$$

$$\times\,(4kp)^{-2it}\frac{\Gamma(1+it)}{\Gamma(1-it)}\frac{F_0\left(t,\frac{\alpha^2}{2}\right)}{\dot{H}_0^+\left(-t,\frac{\chi}{2}\right)}dt, \tag{4.80}$$

$$J_n^{\text{back}} = \frac{-8i\,e^{2ikp-ikp\eta}}{\pi\sqrt{\chi}\sqrt{1-\eta^2}\alpha}\int_{-\infty}^{+\infty}\left(\frac{1+\eta}{1-\eta}\right)^{it}\tilde{Q}_-(t)\times$$

$$\times\,(4kp)^{-2it}\frac{\Gamma\left(\frac{n}{2}+it\right)}{\Gamma\left(\frac{n}{2}-it\right)}\sqrt{\frac{n+2it}{n-2it}}dt. \tag{4.81}$$

For the comparison and validation of the derived asymptotic approximations, we present below the induced currents in the case of TM polarization in two longitudinal sections on the surface: the most shaded at $\varphi = 0$ and the most illuminated at $\varphi = \pi$.

The results are obtained by summing up 10 Fourier harmonics for the forward and backward waves. The integrals (4.78), (4.79), (4.80) and (4.81) are computed by restricting the domain of integration to such interval $[t_1, t_2]$ that for $t < t_1$ and $t > t_2$ it holds that $|f(t)| < 10^{-6} \max |f|$, where f is the integrated function. The integral over $[t_1, t_2]$ is computed as the mean value at equidistant points with the step of integration $\Delta t = 0.025$. It is checked that both the limitation on the number of harmonics and the step size do not affect the 4 leading decimal digits of the result.

Numerically, the currents are represented as the sum of Fourier harmonics with each harmonics computed by the finite-element method. It uses quadrilateral mesh in the domain surrounded by the PML layer. The finite elements are expressed by polynomials of degree 8 which results in a quite accurate approximation. The resulting system of linear equations is solved by a direct solver. More details on the numerical method can be found in [10], where the case of axial incidence was considered. Here, we present the currents induced by a plane wave incident at an angle of 5° to the axis. Compared to [10], we consider more elongated spheroids in order to make the effect of elongation more pronounced. The parameters of the problems are presented in Table 4.1. The results of comparison are shown in Figs. 4.2, 4.3, 4.4, 4.5, 4.6, 4.7 and 4.8. Figures 4.2, 4.3 and 4.4 present the amplitudes of the currents. The results obtained numerically are shown with solid lines and the asymptotic approximations are shown with dashed lines. The numerical and the asymptotic approximations presented in Figs. 4.2 and 4.3 almost coincide except for the values of η close to 1, where the approximation of Sect. 4.5.1 should be used. The agreement for less elongated spheroid shown in Fig. 4.4 is not so accurate. One can conclude that while the mean value is correct the oscillations of the current amplitude are a bit shifted. That means that for less elongated spheroid, the asymptotic approximation for the backward wave becomes inaccurate, while the accuracy of the asymptotic representation for the forward wave remains high.

To see this more evidently, we present in Figs. 4.5, 4.6 and 4.7 the relative error $\delta J = (|J_{\text{num}}| - |J_{\text{as}}|) / |J_{\text{num}}|$. These figures show that at 2 GHz the errors are smaller than at 1 GHz, which is natural for high-frequency asymptotics. However, the derived asymptotic formulas still give a good approximation for the induced currents even at not-too-high frequencies, which is seen from Fig. 4.8, which presents the currents for the case of $kb = 5$ for two spheroids with different elongation. We remark that for smaller kp, one can consider larger incidence angles; Fig. 4.8 corresponds to $\vartheta = 30°$.

It is worth noting that for the very much elongated spheroids, as shown in Figs. 4.5 and 4.6, the error increases at $\eta \to -1$, which means that in this case it is necessary to take into account the effect of backward wave reflection from the front tip of the spheroid which results in the secondary forward wave. By the thin line on the left graphics of Fig. 4.5, we show the relative error for the approximation

$$J = J^{\text{forward}} + J^{\text{back}} + J^{\text{forward}_1},$$

where the current J^{forward_1} of the second-order forward wave is expressed by the integrals (4.78) and (4.79) with the additional multiplier

Table 4.1 Parameters of the test problems for the currents

No.	b (m)	a (m)	f (GHz)	ϑ_0 (°)	kb	χ	α
1	2.50	0.25	1	5	52.396	0.524	0.632
			2	5	104.792	1.048	0.893
2	2.50	0.50	1	5	52.396	2.096	0.632
			2	5	104.792	4.192	0.893
3	2.50	1.00	1	5	52.796	8.383	0.632
			2	5	104.792	16.767	0.893

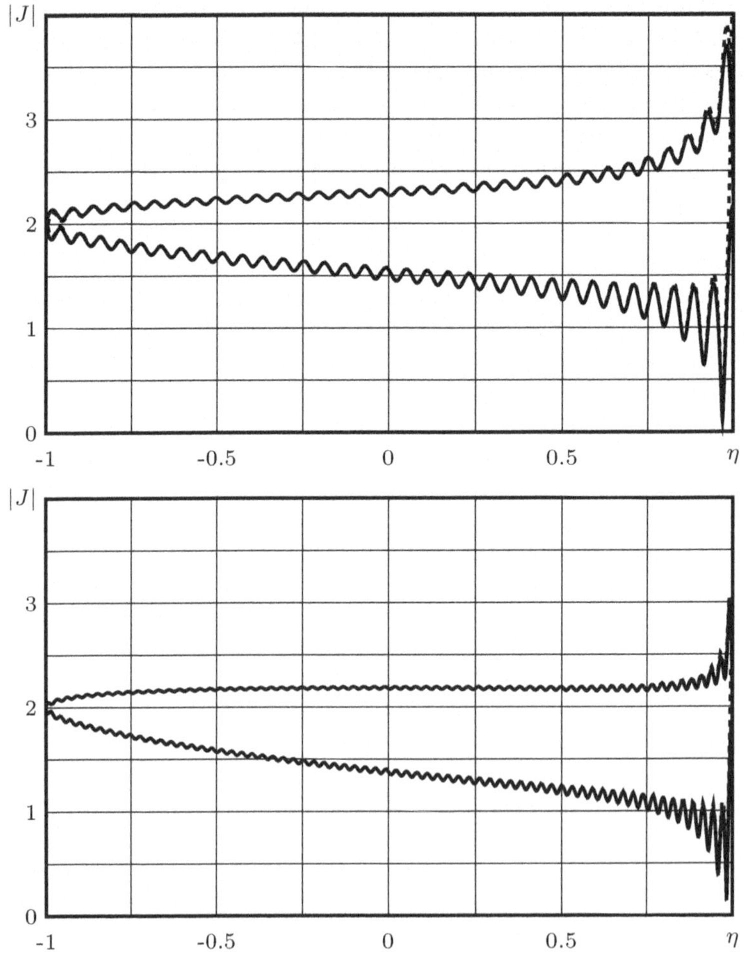

Fig. 4.2 Induced currents at 1 GHz (top) and 2 GHz (bottom) in the case of $b = 2.5$ m, $a = 0.25$ m and $\vartheta_0 = 5°$, TM polarization

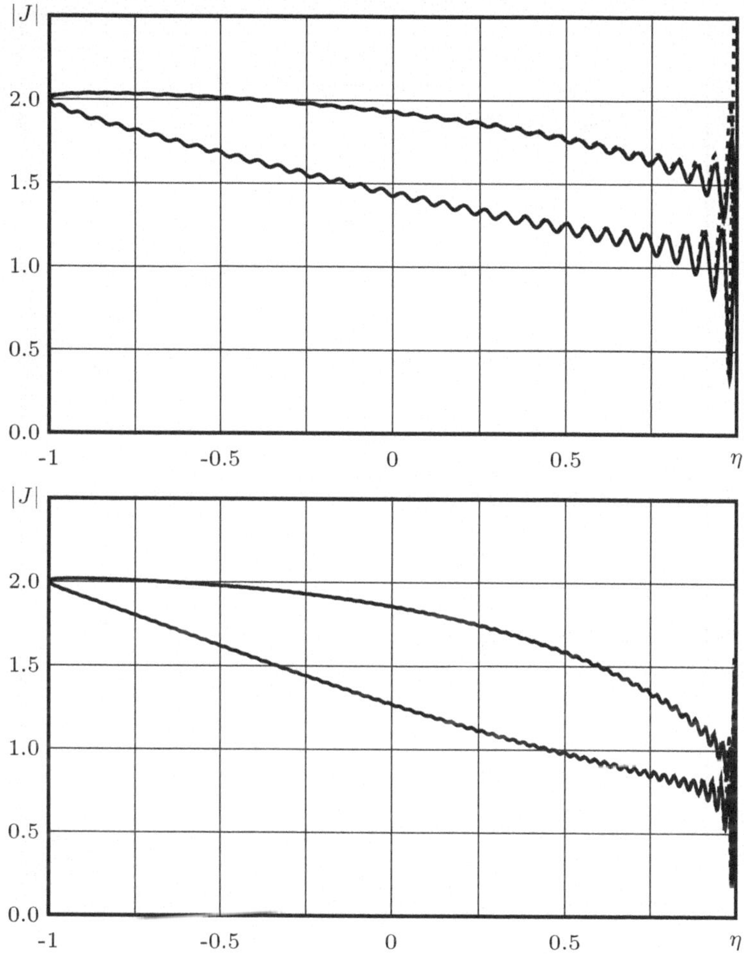

Fig. 4.3 Induced currents at 1 GHz (top) and 2 GHz (bottom) in the case of $b = 2.5$ m, $a = 0.5$ m and $\vartheta_0 = 5°$, TM polarization

$$\mathcal{R}_n = (-1)^{(n-1)} e^{4ikp} (4kp)^{-4it} \frac{\Gamma^2\left(\frac{n}{2} + it\right)}{\Gamma^2\left(\frac{n}{2} - it\right)} \frac{2it + n}{2it - n}$$

under the sign of integration. The multiplier $\mathcal{R}_n$ describes the effect of secondary forward wave excitation by the backward wave, which in its turn is excited by the principal forward wave according to formulas (4.70).

As can be seen in Fig. 4.5, inclusion of this wave improves the agreement with numerical results, especially in the vicinity of the front tip.

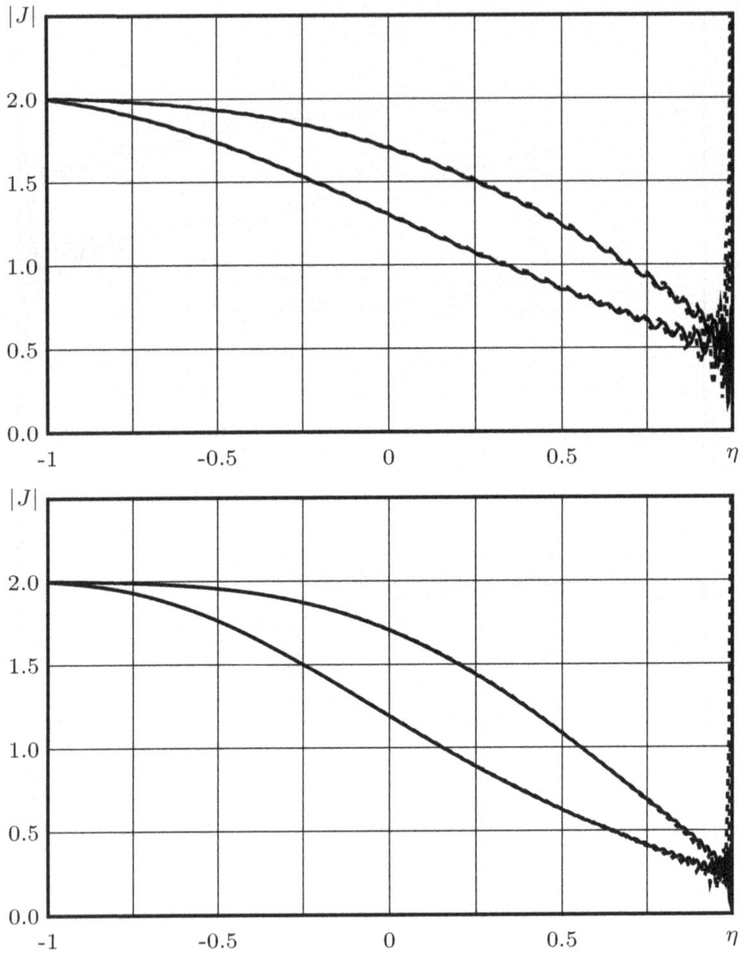

Fig. 4.4 Induced currents at 1 GHz (top) and 2 GHz (bottom) in the case of $b = 2.5$ m, $a = 1.0$ m and $\vartheta_0 = 5°$, TM polarization

4.8 The Far Field

If the currents are known, the Stratton-Chu formula [20] allows the scattered field to be found. Taking into account that the tangential components of the electric vector E are zero on the surface of the perfect conductor, this formula reduces to

$$\mathbf{H}^s = -\frac{1}{4\pi}\iint [\mathbf{J}, \nabla G] dS, \tag{4.82}$$

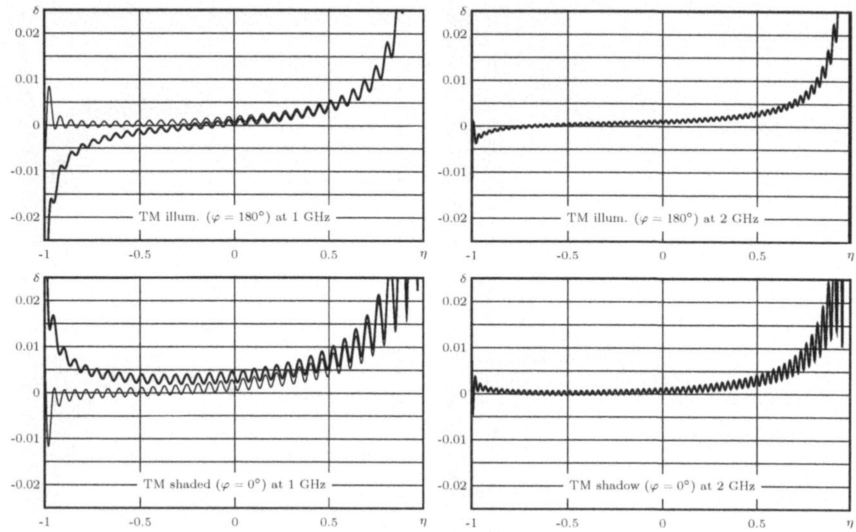

Fig. 4.5 Relative error for spheroid with $b = 2.5$ and $a = 0.25$

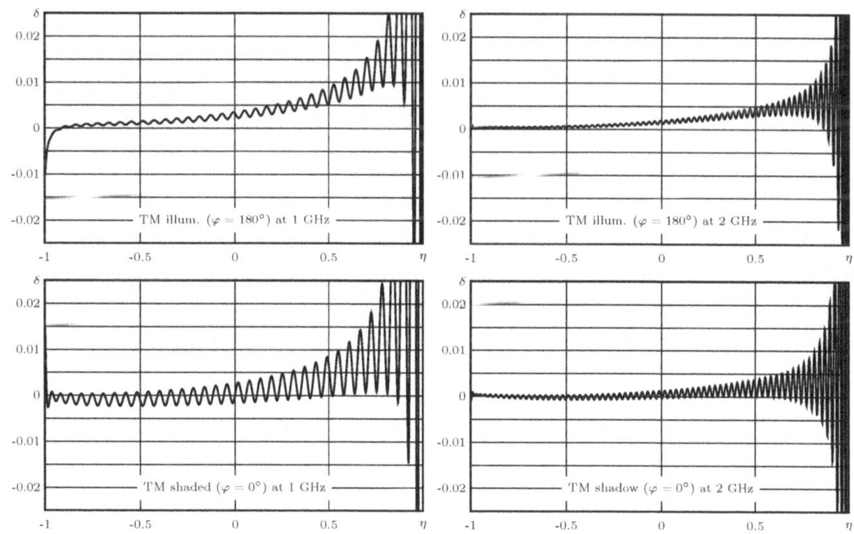

Fig. 4.6 Relative error for spheroid with $b = 2.5$ and $a = 0.5$

where $\mathbf{J}$ is the total induced current, the brackets denote the vector product and G is the scalar Green's function

$$G(\mathbf{r}, \mathbf{r}_0) = \frac{e^{ik|\mathbf{r}-\mathbf{r}_0|}}{|\mathbf{r}-\mathbf{r}_0|}.$$

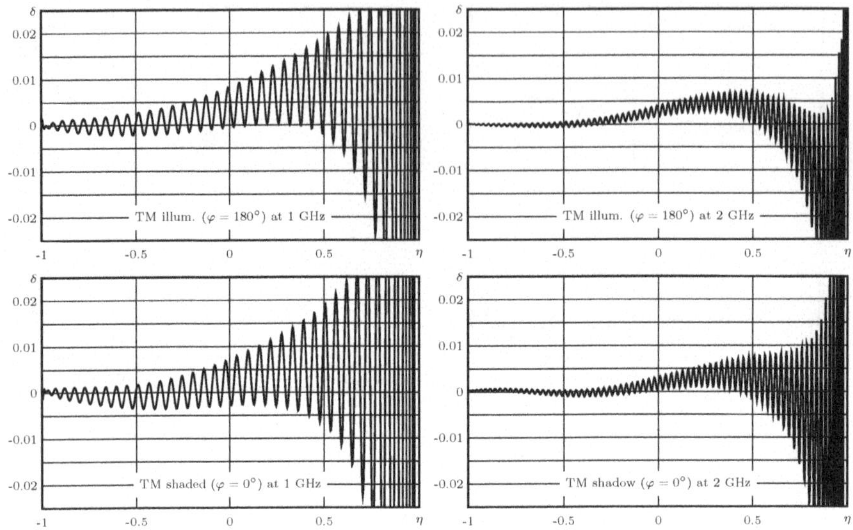

Fig. 4.7 Relative error for spheroid with $b = 2.5$ and $a = 1.0$

Tending the observation point $\mathbf{r}_0$ in (4.82) to infinity along the ray defined by spherical coordinates ϑ and φ, one can calculate the limit under the integration sign and get the formula for the far field amplitude of the magnetic field

$$\boldsymbol{\Psi} = -\frac{1}{4\pi}\iint [\mathbf{J}, \nabla\Psi_G]dS. \tag{4.83}$$

Here, Ψ_G is the far field amplitude of function G. Its asymptotic representation (3.52) as explained in Chap. 3 coincides with the representation of the incident plane wave coming from the opposite direction. It can be written as follows:

$$\Psi_G = \frac{1}{2}\psi_0 + \sum_{m=1}^{+\infty} \psi_m \cos[m(\varphi - \varphi_0)], \tag{4.84}$$

$$\psi_m = \frac{2\mathrm{i}^n \mathrm{e}^{-ikp\eta}}{\pi\sqrt{1-\eta^2}\sqrt{\chi\tau}\beta} \int_{-\infty}^{+\infty} \left(\frac{1+\eta}{1-\eta}\right)^{\mathrm{i}s} \Omega_m(s)\times$$
$$\times M_{\mathrm{i}t,m/2}(\mathrm{i}\beta^2) M_{\mathrm{i}s,\frac{m}{2}}(-\mathrm{i}\chi\tau)\mathrm{d}s,$$

where $\beta = \sqrt{kb\vartheta}$ is the scaled observation angle.

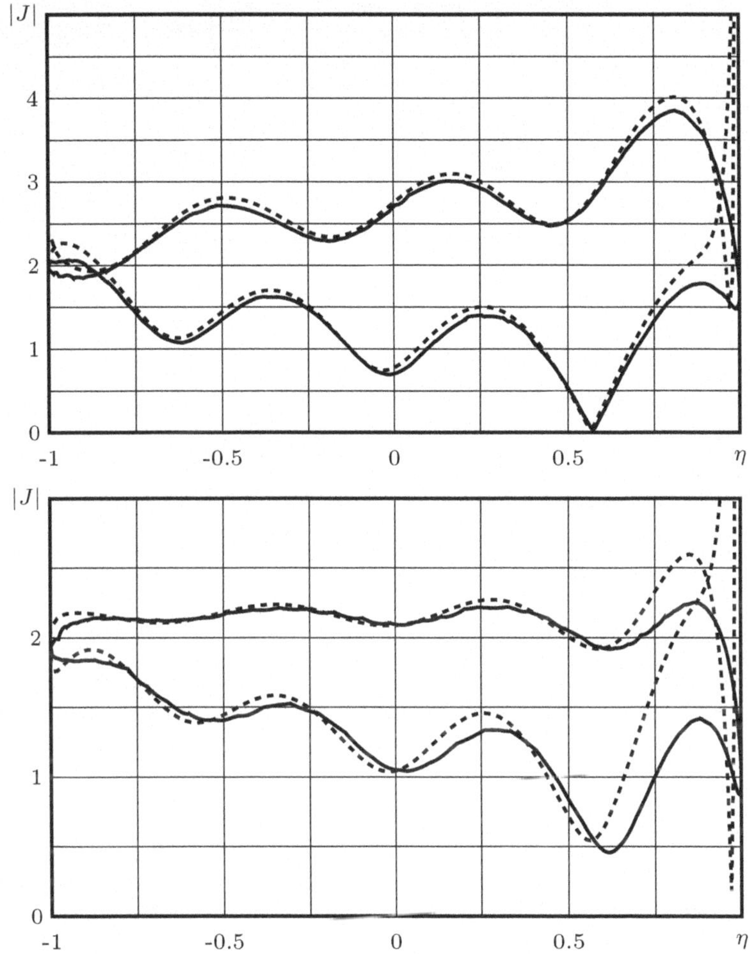

Fig. 4.8 Currents on the spheroid with $kb = 5$, $a : b = 1 : 4$ (top) and $a : b = 1 : 2$ (bottom) for $\vartheta_0 = 30°$, TM polarization

We are going to find the asymptotic representation for the far field amplitude $\boldsymbol{\Psi}$ at the leading order with respect to the asymptotic parameter kp. For that, we substitute in (4.83) the asymptotics of the currents presented in the previous section.

It is more convenient to use Cartesian components of the vectors, but perform the integration in the coordinates of the boundary layer. In particular, we define

$$\mathrm{d}S = b\, d\eta\, a\sqrt{1-\eta^2}\, \mathrm{d}\varphi.$$

Formula (4.83) in the Cartesian components reads

$$\Psi_x = -\frac{1}{4\pi}\iint\left(J_y\frac{\partial\Psi_G}{\partial z} - J_z\frac{\partial\Psi_G}{\partial y}\right)\mathrm{d}S, \tag{4.85}$$

$$\Psi_y = -\frac{1}{4\pi}\iint\left(J_z\frac{\partial\Psi_G}{\partial x} - J_x\frac{\partial\Psi_G}{\partial z}\right)\mathrm{d}S.$$

The unit vector along the η coordinate almost coincides with the unit vector along z, that is $J_z = J_\eta$ at the leading order, but the correction produces the component along r equal to

$$J_r = -\frac{a}{b}\frac{\eta}{\sqrt{1-\eta^2}}J_\eta.$$

In view of $J_\eta = H_\varphi$ and $J_\varphi = -H_\eta$, it is easy to get

$$J_r = -\frac{a}{b}\frac{\eta}{\sqrt{1-\eta^2}}H_\varphi, \quad J_\varphi = \frac{i}{ka\sqrt{1-\eta^2}}\left(2\frac{\partial E_\varphi}{\partial\tau} - \frac{\partial H_\varphi}{\partial\varphi}\right)$$

with Cartesian components defined by the formulas

$$J_x = J_r\cos\varphi - J_\varphi\sin\varphi, \qquad J_y = J_r\sin\varphi + J_\varphi\cos\varphi.$$

Consider now the gradient of Ψ_G. The differentiation with respect to z at the leading order reduces to the multiplication by $-ik$, and the other derivatives can be expressed by the formulas

$$\frac{\partial\Psi_G}{\partial x} = \frac{1}{a\sqrt{1-\eta^2}}\left(\left(2\frac{\partial\Psi_G}{\partial\tau} - \mathrm{i}\chi\eta\Psi_G\right)\cos\varphi - \frac{\partial\Psi_G}{\partial\varphi}\sin\varphi\right),$$

$$\frac{\partial\Psi_G}{\partial y} = \frac{1}{a\sqrt{1-\eta^2}}\left(\left(2\frac{\partial\Psi_G}{\partial\tau} - \mathrm{i}\chi\eta\Psi_G\right)\sin\varphi + \frac{\partial\Psi_G}{\partial\varphi}\cos\varphi\right).$$

We substitute the above expressions into formula (4.85) and get

$$\Psi_x = -\frac{b}{4\pi}\int_{-1}^{1}\mathrm{d}\eta\int_0^{2\pi}\mathrm{d}\varphi\Big\{\left(2\frac{\partial E_\varphi}{\partial\tau} - \frac{\partial H_\varphi}{\partial\varphi}\right)\Psi_G\cos\varphi$$
$$-H_\varphi\left(2\frac{\partial\Psi_G}{\partial\tau}\sin\varphi + \frac{\partial\Psi_G}{\partial\varphi}\cos\varphi\right)\Big\},$$

$$\Psi_y = -\frac{b}{4\pi}\int\limits_{-1}^{1} \mathrm{d}\eta \int\limits_{0}^{2\pi} \mathrm{d}\varphi \Big\{ \Big(2\frac{\partial E_\varphi}{\partial \tau} - \frac{\partial H_\varphi}{\partial \varphi} \Big) \Psi_G \sin\varphi$$

$$+ H_\varphi \Big(2\frac{\partial \Psi_G}{\partial \tau} \cos\varphi - \frac{\partial \Psi_G}{\partial \varphi} \sin\varphi \Big) \Big\}.$$

Now we need to simplify the above expressions. Consider the integrals with respect to the angle φ. The field depends on φ as expressed by formulas (4.14) or (4.28) and (4.84). Therefore, the integrated expressions contain the products of three trigonometric functions. Evidently, nonzero contributions are due to the terms containing the products of three cosines or the products of two sines and one cosine with $|m-n|=1$. One can show that in agreement with the symmetry of the problem, the following formulas hold:

$$\Psi_x^{TE} = \sum_{n=0}^{+\infty} \Psi_{nx}^{TE} \cos(n\varphi_0), \qquad \Psi_y^{TE} = \sum_{n=1}^{+\infty} \Psi_{ny}^{TE} \sin(n\varphi_0),$$

$$\Psi_x^{TM} = \sum_{n=1}^{+\infty} \Psi_{nx}^{TM} \sin(n\varphi_0), \qquad \Psi_y^{TM} = \sum_{n=0}^{+\infty} \Psi_{ny}^{TM} \cos(n\varphi_0).$$

For the harmonics, we get

$$\Psi_{nx}^{TE} = -\frac{b}{4}\int\limits_{-1}^{1} \{\psi_n \mathcal{D}(E_{n+1} + E_{n-1}) - (H_{n+1} - H_{n-1})\mathcal{D}\psi_n\}\,\mathrm{d}\eta,$$

$$\Psi_{ny}^{TE} = -\frac{b}{4}\int\limits_{-1}^{1} \{\psi_n \mathcal{D}(E_{n+1} - E_{n-1}) - (H_{n+1} + H_{n-1})\mathcal{D}\psi_n\}\,\mathrm{d}\eta,$$

$$\Psi_{nx}^{TM} = -\frac{b}{4}\int\limits_{-1}^{1} \{\psi_n \mathcal{D}(E_{n+1} + E_{n-1}) + (H_{n+1} - H_{n-1})\mathcal{D}\psi_n\}\,\mathrm{d}\eta,$$

$$\Psi_{ny}^{TM} = -\frac{b}{4}\int\limits_{-1}^{1} \big\{\psi_n \mathcal{D}(E_{n+1} - E_{n-1}) + \big(H_{n+1}^{+} H_{n-1}\big)\mathcal{D}\psi_n\big\}\,\mathrm{d}\eta,$$

where $\mathcal{D} = \frac{\partial}{\partial\tau} + \frac{1}{2}$. When substituting expressions for the Fourier harmonics E_ℓ and H_ℓ into these formulas, the terms containing Whittaker functions M and W with the second indices equal to $\frac{n-2}{2}$, $\frac{n}{2}$ and $\frac{n+1}{2}$ appear. With the help of the boundary conditions (using properties of the reflection coefficients), one can exclude functions with indices $\frac{n-2}{2}$ and $\frac{n+2}{2}$, which yields the formulas

$$\Psi_{nx}^{TE} = \frac{b\chi}{2} \mathrm{i}^{n+1} \int_{-1}^{1} \{\psi_n \mathcal{D}(A_n + B_n) - (A_n + B_n)\mathcal{D}\psi_n\} \,\mathrm{d}\eta, \tag{4.86}$$

$$\Psi_{ny}^{TE} = \frac{b\chi}{4} \mathrm{i}^{n-1} \int_{-1}^{1} \{\psi_n \mathcal{D}(A_n - B_n) - (A_n - B_n)\mathcal{D}\psi_n\} \,\mathrm{d}\eta.$$

$$\Psi_{nx}^{TM} = \frac{b\chi}{2} \mathrm{i}^{n+1} \int_{-1}^{1} \{\psi_n \mathcal{D}(A_n - B_n) - (A_n - B_n)\mathcal{D}\psi_n\} \,\mathrm{d}\eta,$$

$$\Psi_{ny}^{TM} = \frac{b\chi}{4} \mathrm{i}^{n-1} \int_{-1}^{1} \{\psi_n \mathcal{D}(A_n + B_n) - (A_n + B_n)\mathcal{D}\psi_n\} \,\mathrm{d}\eta. \tag{4.87}$$

Further, we substitute here expressions (4.15) and (4.16) and change the order of integration. The integral with respect to the boundary-layer variable η gives delta-function

$$\int_{-1}^{1} \left(\frac{1-\eta}{1+\eta}\right)^{\mathrm{i}(t-s)} \frac{\mathrm{d}\eta}{1-\eta^2} = \pi\delta(t-s).$$

This eliminates the integration with respect to s and also causes the compensation of the terms containing the Whittaker functions $M_{\mathrm{i}t,n/2}(-\mathrm{i}\chi)$. This confirms the well-known fact that the incident field gives no contribution to $\boldsymbol{\Psi}$. In the other terms, the Wronskian of the Whittaker functions $M_{\mathrm{i}t,n/2}$ and $W_{\mathrm{i}t,n/2}$ can be separated, which simplifies formulas (4.86)–(4.87). In the TE case, we get

$$\Psi_{nx}^{TE} = \frac{ib}{\pi\alpha\beta} \int_{-\infty}^{+\infty} \Omega_n^2(t) M_{\mathrm{i}t,n/2}(\mathrm{i}\beta^2) M_{\mathrm{i}t,n/2}(\mathrm{i}\alpha^2) \times$$
$$\times \Big(R_n(t) + T_n(t)\Big) \frac{(-1)^{n+1}\Gamma(n+1)}{\Gamma\left(\frac{n+1}{2} - \mathrm{i}t\right)} \mathrm{d}t,$$

$$\Psi_{ny}^{TE} = \frac{ib}{\pi\alpha\beta}\int_{-\infty}^{+\infty}\Omega_n^2(t)M_{\mathrm{i}t,n/2}(\mathrm{i}\beta^2)M_{\mathrm{i}t,n/2}(\mathrm{i}\alpha^2)\times$$

$$\times\Big(R_n(t)-T_n(t)\Big)\frac{(-1)^{n+1}\Gamma(n+1)}{\Gamma\left(\frac{n+1}{2}-\mathrm{i}t\right)}\mathrm{d}t,$$

with R_n and T_n defined by the formulas

$$R_1 = -\frac{M_{\mathrm{i}t,1/2}(-\mathrm{i}\chi)}{W_{\mathrm{i}t,1/2}(-\mathrm{i}\chi)},$$

$$R_n = -\Big\{M_{\mathrm{i}t,n/2}(-\mathrm{i}\chi)\dot{W}_{\mathrm{i}t,n/2-1}(-\mathrm{i}\chi)+\dot{M}_{\mathrm{i}t,n/2}(-\mathrm{i}\chi)W_{\mathrm{i}t,n/2-1}(-\mathrm{i}\chi)$$
$$+\frac{\Gamma(n-1)}{\Gamma(\frac{n-1}{2}-\mathrm{i}t)}\frac{\Omega_{n-2}}{\Omega_n}\frac{M_{\mathrm{i}t,n/2-1}(\mathrm{i}\alpha^2)}{M_{\mathrm{i}t,n/2}(\mathrm{i}\alpha^2)}\Big\}\frac{1}{Z_{n-1}},\qquad n=2,3,\ldots$$

$$T_n = -\Big\{M_{\mathrm{i}t,n/2}(-\mathrm{i}\chi)\dot{W}_{\mathrm{i}t,n/2+1}(-\mathrm{i}\chi)+\dot{M}_{\mathrm{i}t,n/2}(-\mathrm{i}\chi)W_{\mathrm{i}t,n/2+1}(-\mathrm{i}\chi)$$
$$+\frac{\Gamma(n+3)}{\Gamma(\frac{n+3}{2}-\mathrm{i}t)}\frac{\Omega_{n+2}}{\Omega_n}\frac{M_{\mathrm{i}t,n/2+1}(\mathrm{i}\alpha)}{M_{\mathrm{i}t,n/2}(\mathrm{i}\alpha)}\Big\}\frac{1}{Z_{n+1}(t)},\qquad n=0,1,2,\ldots$$

In the TM case, the expressions are

$$\Psi_{nx}^{TM} = \frac{ib}{\pi\alpha\beta}\int_{-\infty}^{+\infty}\Omega_n^2(t)M_{\mathrm{i}t,n/2}(\mathrm{i}\beta^2)M_{\mathrm{i}t,n/2}(\mathrm{i}\alpha^2)\times$$

$$\times\Big(T_n(t)-R_n(t)\Big)\frac{(-1)^{n+1}\Gamma(n+1)}{\Gamma\left(\frac{n+1}{2}-\mathrm{i}t\right)}\mathrm{d}t,$$

$$\Psi_{ny}^{TM} = -\frac{ib}{\pi\alpha\beta}\int_{-\infty}^{+\infty}\Omega_n^2(t)M_{\mathrm{i}t,n/2}(\mathrm{i}\beta^2)M_{\mathrm{i}t,n/2}(\mathrm{i}\alpha^2)\times$$

$$\times\Big(T_n(t)+R_n(t)\Big)\frac{(-1)^{n+1}\Gamma(n+1)}{\Gamma\left(\frac{n+1}{2}-\mathrm{i}t\right)}\mathrm{d}t,$$

with R_n and T_n defined by the formulas

$$R_1 = -\frac{\dot{M}_{\mathrm{i}t,1/2}(-\mathrm{i}\chi)}{\dot{W}_{\mathrm{i}t,1/2}(-\mathrm{i}\chi)},$$

Table 4.2 Parameters of the test examples for the far field

	a (m)	b (m)	χ	kb	β for 5°
a)	1.0	2.50	8.383	52.39	0.6316
b)	0.5	1.25	4.192	26.20	0.4467
c)	0.5	1.77	2.960	37.10	0.5315
d)	0.3125	1.39	1.472	29.13	0.4710

$$R_n = -\Big\{M_{\mathrm{i}t,n/2}(-\mathrm{i}\chi)\dot{W}_{\mathrm{i}t,n/2-1}(-\mathrm{i}\chi) + \dot{M}_{\mathrm{i}t,n/2}(-\mathrm{i}\chi)W_{\mathrm{i}t,n/2-1}(-\mathrm{i}\chi)$$

$$-\frac{\Gamma(n-1)}{\Gamma(\frac{n-1}{2}-\mathrm{i}t)}\frac{\Omega_{n-2}}{\Omega_n}\frac{M_{\mathrm{i}t,n/2-1}(\mathrm{i}\alpha^2)}{M_{\mathrm{i}t,n/2}(\mathrm{i}\alpha^2)}\Big\}\frac{1}{Z_{n-1}}, \qquad n = 2, 3, \ldots$$

$$T_n = -\Big\{M_{\mathrm{i}t,n/2}(-\mathrm{i}\chi)\dot{W}_{\mathrm{i}t,n/2+1}(-\mathrm{i}\chi) + \dot{M}_{\mathrm{i}t,n/2}(-\mathrm{i}\chi)W_{\mathrm{i}t,n/2+1}(-\mathrm{i}\chi)$$

$$-\frac{\Gamma(n+3)}{\Gamma(\frac{n+3}{2}-\mathrm{i}t)}\frac{\Omega_{n+2}}{\Omega_n}\frac{M_{\mathrm{i}t,n/2+1}(\mathrm{i}\alpha^2)}{M_{\mathrm{i}t,n/2}(\mathrm{i}\alpha^2)}\Big\}\frac{1}{Z_{n+1}}, \qquad n = 0, 1, 2, \ldots$$

The denominators Z_n in both TE and TM cases are given by expressions (4.24) or equivalently by (4.25).

4.9 Test Examples

In this section, we present a test comparison of the asymptotic and numerical results for the far field amplitudes. The parameters of the test problems are presented in Table 4.2.

The far fields are expressed by formulas of the previous section in the form of the integrals containing Whittaker functions. To use the code developed in [21], we rewrite these formulas by replacing Whittaker functions with Coulomb wave functions. These cumbersome, though straightforward, derivations result in the following final expressions for the far field amplitudes of the magnetic vector. In the case of TE polarization, we get

$$\Psi_x^{TE} = \frac{8ib}{\pi\alpha\beta}\sum_{\ell=0}^{+\infty}\int_{-\infty}^{+\infty}(a_\ell^+ + b_\ell^+)\cos(\ell\varphi)\mathrm{d}t, \tag{4.88}$$

$$\Psi_y^{TE} = \frac{8ib}{\pi\alpha\beta}\sum_{\ell=0}^{+\infty}\int_{-\infty}^{+\infty}(a_\ell^+ - b_\ell^+)\sin(\ell\varphi)\mathrm{d}t$$

and in the case of TM polarizations, the expressions are

$$\Psi_x^{TM} = \frac{8ib}{\pi\alpha\beta}\sum_{\ell=0}^{+\infty}\int_{-\infty}^{+\infty}(a_\ell^- + b_\ell^-)\sin(\ell\varphi)dt,$$

$$\Psi_y^{TM} = \frac{8ib}{\pi\alpha\beta}\sum_{\ell=0}^{+\infty}\int_{-\infty}^{+\infty}(a_\ell^- - b_\ell^-)\cos(\ell\varphi)dt. \tag{4.89}$$

In the above formulas, the following notations are introduced

$$a_\ell^\pm = \frac{F_{\frac{\ell-1}{2}}\left(t, \frac{1}{2}\alpha^2\right)}{z_\ell}\left(y_\ell F_{\frac{\ell-1}{2}}\left(t, \frac{1}{2}\beta^2\right) \pm F_{\frac{\ell+1}{2}}\left(t, \frac{1}{2}\beta^2\right)\right), \tag{4.90}$$

$$b_0^+ = 0, \quad b_1^+ = F_0\left(t, \tfrac{1}{2}\alpha^2\right) F_0\left(t, \tfrac{1}{2}\beta^2\right)\frac{F_0\left(-t, \frac{1}{2}\chi\right)}{H_0^+\left(-t, \frac{1}{2}\chi\right)}, \tag{4.91}$$

$$b_0^- = 0, \quad b_1^- = F_0\left(t, \tfrac{1}{2}\alpha^2\right) F_0\left(t, \tfrac{1}{2}\beta^2\right)\frac{\dot F_0\left(-t, \frac{1}{2}\chi\right)}{\dot H_0^+\left(-t, \frac{1}{2}\chi\right)}, \tag{4.92}$$

$$b_{\ell+2}^\pm = \frac{F_{\frac{\ell+1}{2}}\left(t, \frac{1}{2}\alpha^2\right)}{z_\ell}\left(F_{\frac{\ell-1}{2}}\left(t, \frac{1}{2}\beta^2\right) \pm y_\ell F_{\frac{\ell+1}{2}}\left(t, \frac{1}{2}\beta^2\right)\right), \tag{4.93}$$

$$y_\ell = H_{\frac{\ell-1}{2}}^+\left(-t, \tfrac{1}{2}\chi\right)\dot F_{\frac{\ell+1}{2}}\left(-t, \tfrac{1}{2}\chi\right) + \dot H_{\frac{\ell-1}{2}}^+\left(-t, \tfrac{1}{2}\chi\right)F_{\frac{\ell+1}{2}}\left(-t, \tfrac{1}{2}\chi\right), \tag{4.94}$$

$$z_\ell = H_{\frac{\ell-1}{2}}^+\left(-t, \tfrac{1}{2}\chi\right)\dot H_{\frac{\ell+1}{2}}^+\left(-t, \tfrac{1}{2}\chi\right) + \dot H_{\frac{\ell-1}{2}}^+\left(-t, \tfrac{1}{2}\chi\right)H_{\frac{\ell+1}{2}}^+\left(-t, \tfrac{1}{2}\chi\right). \tag{4.95}$$

Here by the dots, we denote the derivatives of the Coulomb wave functions with respect to their second argument. For the case of $\alpha = 0$ or $\beta = 0$, the representations (4.88)–(4.89) contain ambiguities. When avoiding them, we take into account the fact that

$$\frac{F_{-1/2}\left(t, \frac{1}{2}\beta^2\right)}{\beta} \to \sqrt{\frac{\pi/2}{e^{2\pi i t} + 1}} \tag{4.96}$$

while the terms containing Coulomb wave functions F_ℓ with other indices vanish.

Though formulas (4.88)–(4.95) are rather cumbersome and contain infinite integration and summation, computational difficulties are mostly encountered in the computation of the Coulomb wave functions, but are effectively resolved by using the FORTRAN program developed in [21]. The integrals in formulas (4.88)–(4.89) exponentially converge at infinity. For the purpose of numerical computation, we discard the semi-infinite intervals of the integration variable t on which the absolute value of the integrated function is more than a million times less than its maximal

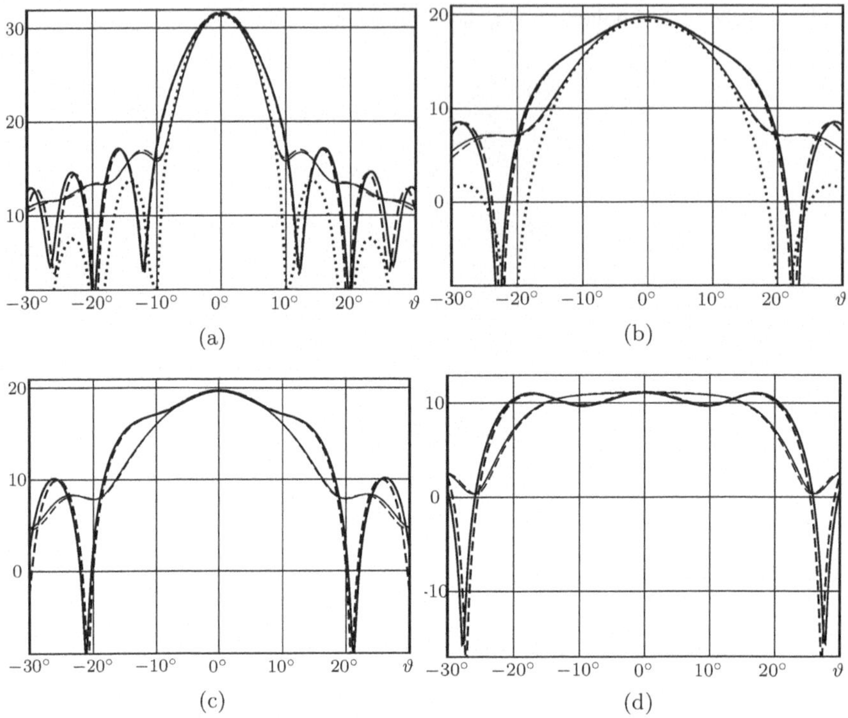

Fig. 4.9 The RCS of the four spheroids for the axial incidence

value. For the remaining finite interval, it is usually sufficient to choose the step of integration equal to 0.4. To minimize the computational errors, we used a 10 times smaller step, i.e. 0.04, which typically results in 150–300 integration nodes, less for smaller values of the elongation parameter χ and scaled angles α and β. The number of terms in the series depends on the parameters α and β, but it is usually sufficient to consider no more than 10 Fourier harmonics. We used 20 terms for the results presented below.

Numerically, the values of the radar cross-sections (RCS) were computed by the method of moments with the help of ANSYS HFSS. The surface of each spheroid was approximated with 142626–14400 triangles. More details can be found in [15]. The results for the case of the axial incidence are presented in Fig. 4.9 and for the incidence at $\vartheta_0 = 5°$, in the case of TE polarization in Fig. 4.10 and in the case of the TM polarization in Fig. 4.11. The values obtained numerically with the use of ANSYS are plotted with the solid lines, and the asymptotic results are plotted with the dashed lines. More thick lines correspond to the fields at the section $\varphi = 0°$ and more thin lines are for the section at $\varphi = 90°$.

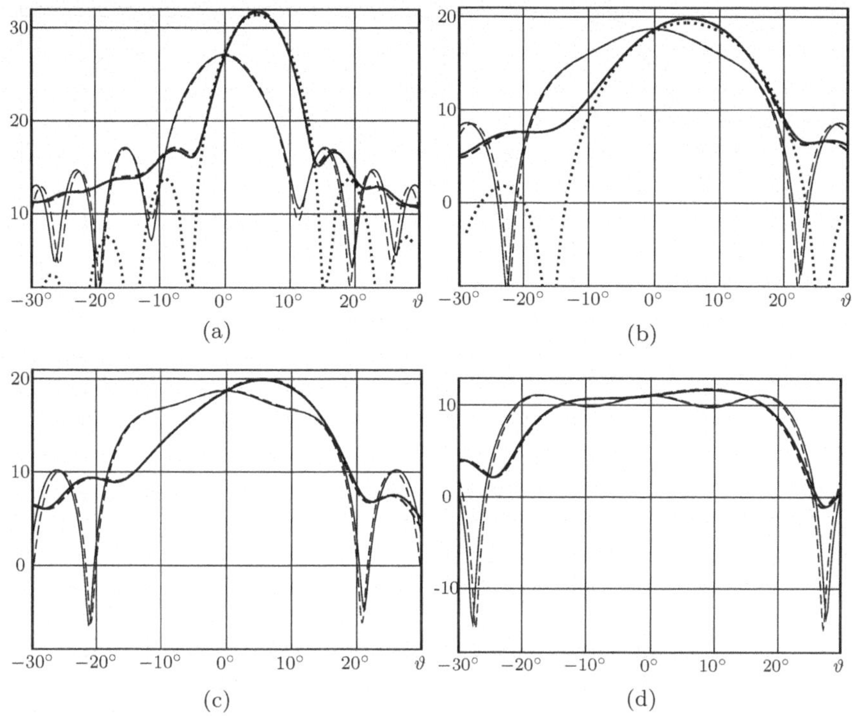

Fig. 4.10 The RCS of the four spheroids for the TE wave incident at 5° to the axis

The test example (a) was the most difficult for ANSYS, and we see some violation of the symmetry of the RCS computed by ANSYS in Figs. 4.10a and 4.11a. All the other test examples demonstrate sufficiently good agreement between the numerical and the asymptotic results.

Figures 4.9 through 4.11 show also Kirchhoff approximation for test examples (a) and (b) with dotted curves. For the tests (c) and (d), we do not show the results of the Kirchhoff approximation, since its accuracy is low.

For the ten times lower frequency (100 MHz), the computations for ANSYS are evidently easier and more accurate. The asymptotic results on the contrary contain larger errors because the asymptotic parameter kp is not large enough. The results of comparison for 100 MHz at axial incidence are presented in Fig. 4.12, which shows that even for those tests the asymptotic approximations (4.88)–(4.89) provide relatively accurate results near the axis. Moreover, the agreement is better for the more elongated spheroids (c) and (d) although the asymptotic parameter kb is larger for the spheroid (a).

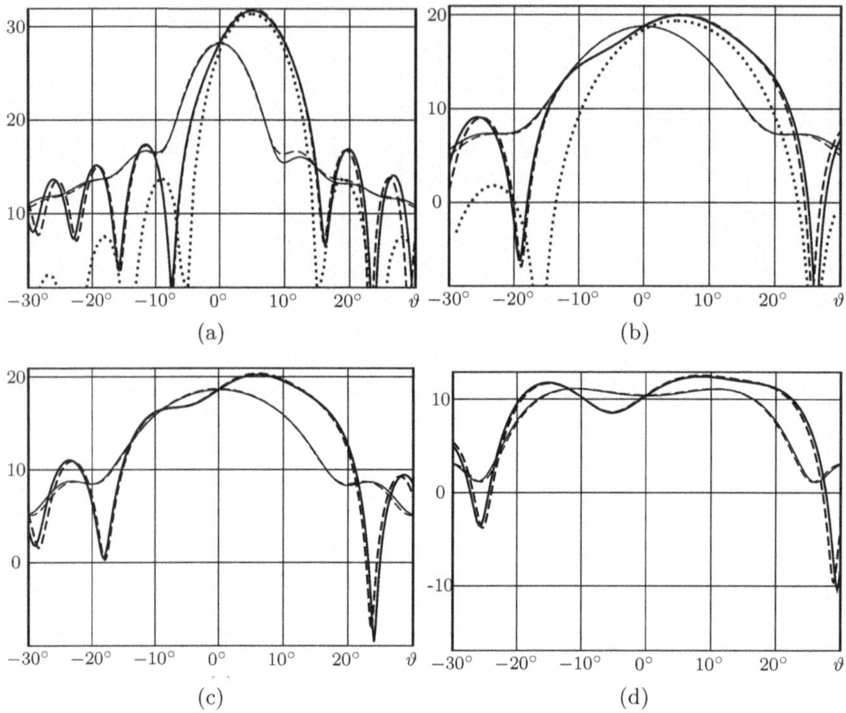

Fig. 4.11 The RCS of the four spheroids for the TM wave incident at 5° to the axis

Having an accurate approximation for the far fields, one can examine specific effects of diffraction by strongly elongated bodies. Figures 4.13 and 4.14 show the RCS in a 30° cone for plane wave incident at the angle of 5° to the axis of spheroids with different rates of elongation. These figures show an effect similar to that in the case of the acoustic wave scattering. Namely, if the body is not too much elongated, the main beam of the RCS is the shadow beam oriented along the direction of incidence. When the rate of elongation increases this beam becomes less pronounced. In the TE case, the magnitude of RCS at the mirror direction of $\varphi = 90°$ increases, and this direction becomes the principal direction of scattering. Simultaneously, the two spots with the minimal intensity of the scattered field appear. In the TM case, the shadow beam spreads with respect to the revolution angle φ and the RCS becomes almost symmetric with respect to φ. At the center, a kind of Arago spot is formed.

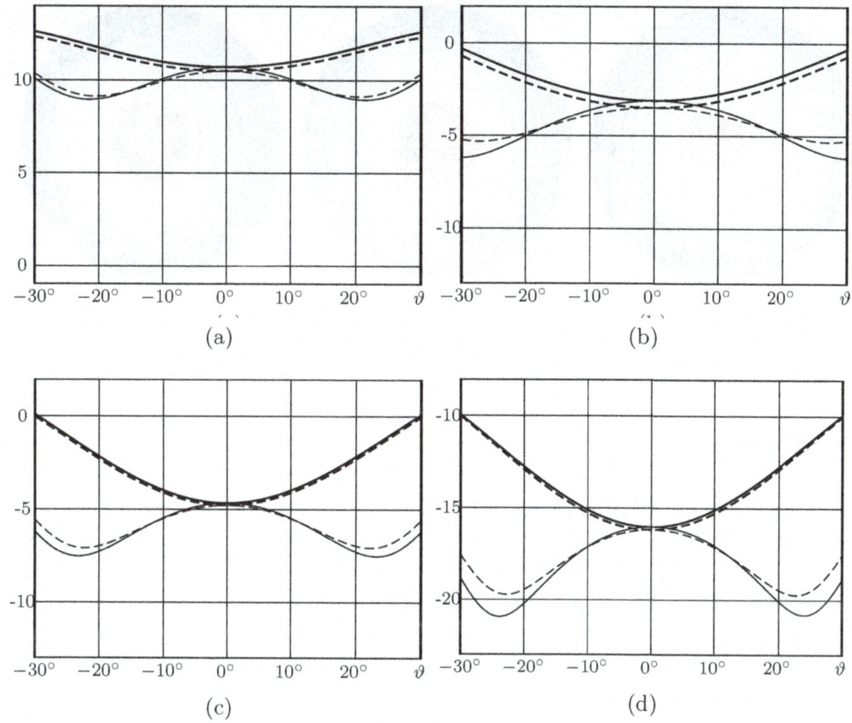

Fig. 4.12 The RCS of the four spheroids at 0.1 GHz

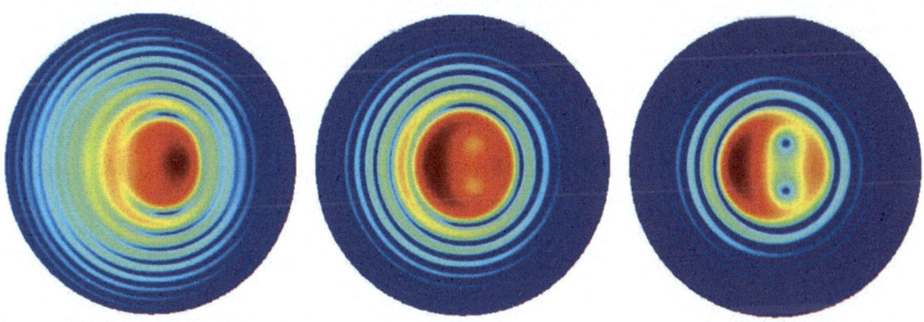

Fig. 4.13 The RCS in a 30° cone of the TE plane wave at 1 GHz incident at 5° on spheroids with $b = 10$ m and $a = 1$ m (left), $a = 0.5$ m (center) and $a = 0.2$ m (right)

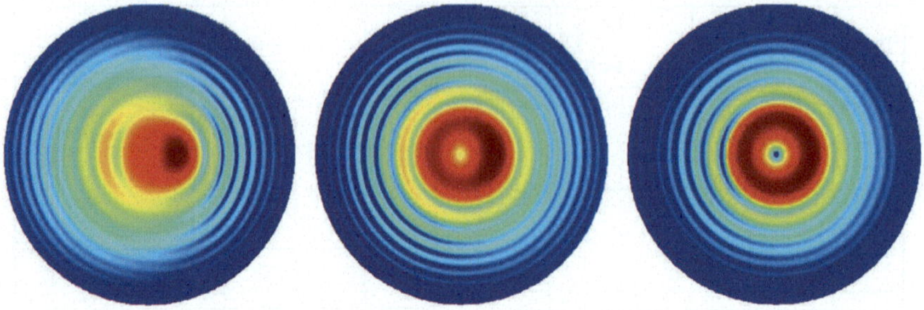

Fig. 4.14 The RCS in a 30° cone of the TM plane wave at 1 GHz incident at 5° on spheroids with $b = 10$ m and $a = 1$ m (left), $a = 0.5$ m (center) and $a = 0.2$ m (right)

References

1. M. Abramowitz, I.A. Stegun, *Handbook of Mathematical Functions with Formulas, Graphs and Mathematical Tables*, National Bureau of Standards, Applied Mathematics Series 55 (1964)
2. I.V. Andronov, High frequency asymptotics of electromagnetic field on a strongly elongated spheroid. PIERS Online **5**(6), 536–540 (2009)
3. I.V. Andronov, High-frequency asymptotics for diffraction by a strongly elongated body. Antennas Wirel. Propag. Lett. IEEE **8**, 872 (2009)
4. I.V. Andronov, Diffraction of a high-frequency electromagnetic wave by an elongated body of revolution. J. Commun. Technol. Electron. **56**(11), 1317–1323 (2011)
5. I.V. Andronov, The currents induced by a high-frequency wave incident at a small angle to the axis of strongly elongated spheroid. PIER M **28**, 273–287 (2013)
6. I.V. Andronov, High-frequency asymptotics for the total scattering cross-section of strongly elongated spheroids. Antennas Wirel. Propag. Lett. **12**, 304 (2013)
7. I.V. Andronov, Diffraction of a high-frequency electromagnetic wave incident at a small angle to the axis of a strongly elongated body. J. Commun. Technol. Electron. **59**(2), 130–138 (2014)
8. I.V. Andronov, Axial diffraction of a dipole field by a strongly elongated spheroid. J. Electromagn. Waves Appl. **32**(12), 1535–1540 (2018)
9. I.V. Andronov, D. Bouche, Forward and backward waves in high-frequency diffraction by an elongated spheroid. Prog. Electromagn. Res. B **29**, 209–231 (2011)
10. I.V. Andronov, D. Bouche, M. Duruflé, High-frequency diffraction of plane electromagnetic wave by an elongated spheroid. IEEE AP Trans. **60**(11), 5286–5295 (2012)
11. I.V. Andronov, D. Bouche, M. Duruflé, High-frequency currents on a strongly elongated spheroid. IEEE Trans. Antennas Propag. **65**(2), 794–804 (2017)
12. I.V. Andronov, R. Mittra, High-frequency asymptotics for the radar cross-section computation of a prolate spheroid with high aspect ratio. IEEE Trans. Antennas Propag. **63**(1), 336–343 (2015)
13. I.V. Andronov, R. Mittra, High-frequency diffraction by an elliptic cylinder: the far field. J. Electromagn. Waves Appl. **29**(10), 1317–1328 (2015)
14. I.V. Andronov, R. Mittra, Recent advances in the asymptotic theory of diffraction by elongated bodies (invited paper). PIER **150**, 163–182 (2015)
15. I.V. Andronov, D.A. Shevnin, High-frequency scattering by perfectly conducting prolate spheroids. J. Electromagn. Waves Appl. **28**(11), 1388–1396 (2014)
16. J.J. Bowman, T.B.A. Senior, P.L.E. Uslenghi, *Electromagnetic and Acoustic Scattering by Simple Shapes* (North-Holland Publishing Co., Amsterdam, 1969)
17. V.A. Fock, Theory of diffraction by a paraboloid of revolution (in Russian), in *Diffraction of Electromagnetic Waves by Some Bodies of Revolution* (Soviet Radio, Moscow, 1957), pp. 5–56

18. I.S. Gradshtein, I.M. Ryzhik, *Table of Integrals, Series and Products* (Elsevier, Academic Press, New York, 2007)
19. I.V. Komarov, L.I. Ponomarev, S.Yu. Slavyanov, *Spheroidal and Coulomb Spheroidal Functions, Science* (Moscow, 1976)
20. J.A. Stratton, L.J. Chu, Diffraction theory of electromagnetic waves. Phys. Rev. **56**(1), 99–107 (1939)
21. I.J. Thompson, A.R. Barnett, COULCC: a continued-fraction algorithm for Coulomb functions of complex order with complex arguments. Comput. Phys. Commun. **36**, 363–372 (1985)

Chapter 5
Other Strongly Elongated Shapes

5.1 Introduction

The asymptotic representations considered in the previous chapters are given in the form of integrals which express the result of variable separation. Though only the leading order approximation is considered, it can be checked that at any asymptotic order the amplitudes which appear under the sign of integration can be found explicitly. This is due to the following peculiarities of the developed asymptotic procedure. First, the boundary-layer coordinates associated with the surface of the body lead to the parabolic equation which allows variable separation. The second circumstance is that the dependence on coordinate η which varies along the surface is such that the result of variable separation represented by the integral with respect to the parameter coincides with the integral transform (A.32) which has the inverse (A.33). This allows the corresponding representation for the incident field to be written explicitly. The third successful circumstance is in the form of the integral equation that appears when the integral representation for the attenuation function is substituted into the boundary conditions. This equation can be easily solved by setting the relation between the amplitudes of the incident and scattered/reflected fields. For that, it is important that the boundary conditions considered in the previous chapters are the Dirichlet, the Neumann or the condition of a perfect conductor. It is worth noting that the case of the impedance surface yields more complicated integral equations and up to now the explicit expression for the solution is found only in the case of the strip [5], which is the limiting case of the elliptic cylinder with its minor semiaxis tending to zero.

The above-mentioned specifics of the boundary-value problems are connected to the shape of the surface. In the problems considered in Chaps. 2–4, the shape of the body is formed of the elongated ellipse, which is the cross-section of the elliptic cylinder in Chap. 2 or forms the surface of the spheroid in Chaps. 3 and 4. In this chapter, we start by considering an arbitrary body of revolution and then by requiring the explicit form of the solution to impose the restrictions on it. So, finally

I. Andronov, *Problems of High Frequency Diffraction by Elongated Bodies*, Springer Series in Optical Sciences 243, https://doi.org/10.1007/978-981-99-1276-6_5

we identify only some canonical shapes and present the leading order approximations for the fields of diffraction on these canonical surfaces.

5.2 The Boundary-Layer Coordinates

If the rate of elongation is characterized by the ratio of curvatures ρ_t, ρ in transverse and longitudinal directions, then the above-studied variants of asymptotic expansions correspond to

$$\chi = k\rho\left(\frac{\rho_t}{\rho}\right)^{3/2} = O(1),$$

where k is the wave number.

If χ is asymptotically large, then we deal with the case of diffraction by an ordinary body, and the well-known asymptotic approach described in Chap. 1 can be used for the asymptotic analysis of the diffracted fields. In that case, one deals with the boundary layer that has the sizes of order $(k\rho)^{-1/3}$ in length and of order $(k\rho)^{-2/3}$ in the transverse directions. This boundary layer is referred to as the Fock domain.

If $\chi = O(1)$, then the whole body (for example, the spheroid or the cross-section of the elliptic cylinder) lies in the Fock domain. On the basis of this observation, let us introduce the stretched coordinates (η, ν) on an arbitrary shaped strongly elongated body by the formulas

$$\begin{cases} z = k^{-1/3} f(\eta) + k^{-1} a(\eta)\nu, \\ r = k^{-2/3} g(\eta)\sqrt{\nu}. \end{cases} \tag{5.1}$$

For brevity, we use the decompositions with respect to the powers of the dimensioned parameter k. To exclude dimensionality, one can assume that k is multiplied by some characteristic size of the surface. Function $g(\eta)$ describes the shape of the surface, which is assumed to coincide with the coordinate line $\nu = const$. We consider it to be $\nu = 1$. Functions $f(\eta)$ and $a(\eta)$ are to be chosen later. With the appropriate choice of these functions, one can make the coordinate system (η, ν, φ) orthogonal, at least on the surface $\nu = 1$ and at least at the principal order with respect to asymptotically large k. Besides, and more importantly we require that the parabolic equation which appears for the attenuation function (see below) allows variable separation.

Our nearest goal is to derive the metric tensor of the system introduced by formulas (5.1). Let M denote a point. We compute the derivatives

$$\frac{\partial M}{\partial \eta} = \left[k^{-1/3} f' + k^{-1} a'\nu\right] e_z + k^{-2/3} g'\sqrt{\nu} e_r,$$

$$\frac{\partial M}{\partial \nu} = k^{-1} a e_z + k^{-2/3} \frac{g}{2\sqrt{\nu}} e_r.$$

The orthogonality condition for vectors $\partial M/\partial \eta$ and $\partial M/\partial \nu$ reads

$$k^{-4/3}\left[af' + \frac{gg'}{2}\right] + k^{-2}aa'\nu = 0.$$

We cannot satisfy it exactly, but we can satisfy it in the principal order by setting

$$a = -\frac{1}{2}\frac{gg'}{f'}. \tag{5.2}$$

The metric tensor can be easily found. Its nonzero components are

$$\begin{aligned}
G_{\eta\eta} &= k^{-2/3}(f')^2 + k^{-4/3}\left[2f'a' + (g')^2\right]\nu + k^{-2}(a')^2\nu^2,\\
G_{\eta\nu} &= k^{-2}aa'\nu,\\
G_{\nu\nu} &= k^{-4/3}\frac{g}{4\nu} + k^{-2}a^2,\\
G_{\varphi\varphi} &= k^{-4/3}g^2\nu.
\end{aligned}$$

Further, we need its determinant, which is

$$\begin{aligned}
\det G = k^{-4/3}g^2 \Bigg\{ &\left(k^{-4/3}\frac{g^2}{4} + k^{-2}a^2\nu\right)\left(k^{-2/3}(f')^2 + \right.\\
&\left. +k^{-4/3}\left(2f'a' + (g')^2\right)\nu + k^{-2}(a')^2\nu^2\right) - k^{-4}a^2(a')^2\nu^3\Bigg\},
\end{aligned}$$

$$\det G = k^{-10/3}\frac{1}{4}g^4(f')^2 + k^{-4}\left[g^2a^2(f')^2 + \frac{1}{2}g^4f'a' + \frac{1}{4}g^4(g')^2\right]\nu + \cdots$$

With the use of the expressions for the metric tensor, we rewrite Helmholtz equation in coordinates (η, ν, φ), where φ is the usual polar angle of cylindrical coordinates (r, φ, z). We search for the solution of the Helmholtz equation in the form of the wave, which runs along the surface of the body in the positive direction of z. Using the parabolic equation method, we assume that

$$u = \exp(ikz_0)\sum_{\ell=-\infty}^{+\infty} U_\ell(\eta, \nu)e^{i\ell\varphi},$$

where $z_0 = k^{-1/3}f$ is the value of the z coordinate taken on the axis of the body, i.e. at $\nu = 0$. The harmonics U_ℓ can be considered separately, and below we suppress index ℓ.

One can check that in order to get the principal order equations with respect to $k \gg 1$, it is possible to neglect the extradiagonal terms $G_{\eta\nu}$ in the metric tensor G. Then Helmholtz equation can be written as if the coordinate system is orthogonal

$$\frac{1}{\sqrt{\det G}}\left\{\frac{\partial}{\partial\eta}\left(\frac{\sqrt{\det G}}{G_{\eta\eta}}\frac{\partial}{\partial\eta}\right)+\frac{\partial}{\partial\nu}\left(\frac{\sqrt{\det G}}{G_{\nu\nu}}\frac{\partial}{\partial\nu}\right)\right\}u$$
$$+\frac{1}{G_{\varphi\varphi}}\frac{\partial^2 u}{\partial\varphi^2}+k^2u=0. \qquad (5.3)$$

Here, we have taken into account that tensor G does not depend on polar angle φ. We can rewrite (5.3) as

$$\frac{1}{G_{\eta\eta}}\frac{\partial^2 u}{\partial\eta^2}+\frac{1}{\sqrt{\det G}}\frac{\partial}{\partial\eta}\left(\frac{\sqrt{\det G}}{G_{\eta\eta}}\right)\frac{\partial u}{\partial\eta}+\frac{1}{G_{\nu\nu}}\frac{\partial^2 u}{\partial\nu^2}$$
$$+\frac{1}{\sqrt{\det G}}\frac{\partial}{\partial\nu}\left(\frac{\sqrt{\det G}}{G_{\nu\nu}}\right)\frac{\partial u}{\partial\nu}+\frac{1}{G_{\varphi\varphi}}\frac{\partial^2 u}{\partial\varphi^2}+k^2u=0. \qquad (5.4)$$

Now we compute the terms presented in (5.4)

$$\frac{\partial u}{\partial\eta}\Rightarrow \mathrm{i}k^{2/3}f'U+\frac{\partial U}{\partial\eta},\quad \frac{\partial^2 u}{\partial\eta^2}\Rightarrow -k^{4/3}(f')^2U+2\mathrm{i}k^{2/3}f'\frac{\partial U}{\partial\eta}+\mathrm{i}k^{2/3}f''U+\frac{\partial U}{\partial\eta^2},$$

$$G_{\eta\eta}=k^{-2/3}(f')^2\left[1+k^{-2/3}\frac{2f'a'+(g')^2}{(f')^2}\nu+k^{-4/3}\frac{(a')^2}{(f')^2}\nu^2+\cdots\right],$$

$$\frac{1}{G_{\eta\eta}}=k^{2/3}(f')^{-2}-\frac{2f'a'+(g')^2}{(f')^4}\nu+k^{-2/3}\left[\frac{\left(2f'a'+(g')^2\right)^2}{(f')^6}-\frac{(a')^2}{(f')^4}\right]\nu^2+\cdots$$

$$\frac{\partial}{\partial\nu}\left(\frac{1}{G_{\nu\nu}}\right)=k^{4/3}\frac{4}{g^2}+\cdots$$

Substituting all these expressions in (5.4), it is easy to see that the terms of the order k^2 cancel each other. Equating terms of the order $k^{4/3}$ yields

$$\frac{2f'a'+(g')^2}{(f')^2}\nu U+\frac{2\mathrm{i}}{f'}U_\eta+\mathrm{i}\frac{f''}{(f')^2}U+\frac{\mathrm{i}}{g^2f'}\frac{\partial}{\partial\eta}\left(\frac{g^2f'}{(f')^2}\right)f'U$$
$$+\frac{4\nu}{g^2}\frac{\partial^2 U}{\partial\nu^2}+\frac{4}{g^2}\frac{\partial U}{\partial\nu}-\frac{\ell^2}{g^2\nu}U=0.$$

By multiplying this equation by $g^2/4$, we get it in the form suitable for the variable separation

$$\nu\frac{\partial^2 U}{\partial \nu^2}+\frac{\partial U}{\partial \nu}-\frac{\ell^2}{4\nu}U+\frac{g^2}{4}\frac{2f'a'+(g')^2}{(f')^2}\nu U$$
$$+\,\mathrm{i}\frac{g^2}{2f'}\frac{\partial U}{\partial \eta}+\mathrm{i}\frac{g^2 f''}{4(f')^2}U+\frac{\mathrm{i}}{4}\frac{\partial}{\partial \eta}\left(\frac{g^2}{f'}\right)U=0. \quad (5.5)$$

We see that there is only one problematic term. The last term in the first line of (5.5) may prevent the variable separation. Inorder to make variable separation possible, the coefficient at U in this term should not depend on η. Thus, we require that functions a, f and g satisfy the following condition:

$$\frac{g^2}{4}\left(2\frac{a'}{f'}+\left(\frac{g'}{f'}\right)^2\right)=\frac{C}{4}=const.$$

Substituting here expression (5.2) for a reduces this condition to differential equation

$$\left(\frac{g}{f'}\right)^3\left(g'f''-f'g''\right)=\frac{C}{4}. \quad (5.6)$$

If functions f and g satisfy this equation, then the variables can be separated in the parabolic equation. Namely, particular solutions U of (5.5) can be represented as the product

$$U(\eta,\nu)=F(\eta)G(\nu),$$

where functions F and G satisfy the ordinary differential equations

$$\nu G''+G'-\frac{\ell^2}{4\nu}G-\mu G+\frac{C}{4}\nu G=0, \quad (5.7)$$

$$\mathrm{i}\frac{g^2}{2f'}F'+\mathrm{i}\frac{g^2 f''}{4(f')^2}F+\frac{\mathrm{i}}{4}\frac{\partial}{\partial \eta}\left(\frac{g^2}{f'}\right)F+\mu F=0. \quad (5.8)$$

Here, μ is the parameter of variable separation. Equation (5.7) can be reduced to the Whittaker equation. For that, we represent G as

$$G=\frac{V(\nu)}{\sqrt{\nu}},$$

which excludes the first-order derivative and (5.7) reduces to

$$V''+\left(\frac{C}{4}-\frac{\mu}{\nu}+\frac{1-\ell^2}{4\nu^2}\right)V=0.$$

If $C \neq 0$, then changing the variable

$$\nu = \frac{\mathrm{i}}{\sqrt{C}} t, \qquad V(\nu) = v(t)$$

makes this equation coincident with the Whittaker equation [1]

$$v'' + \left(-\frac{1}{4} - \frac{\mathrm{i}\mu}{\sqrt{C}\nu} + \frac{1-\ell^2}{4\nu^2} \right) v = 0.$$

Thus, in the case of $C \neq 0$, the general solution of the radial (5.7) can be expressed in terms of Whittaker functions M and W with arbitrary coefficients A and B:

$$G = \frac{1}{\sqrt{\nu}} \left\{ A M_{m,\ell/2}\left(-\mathrm{i}\sqrt{C}\nu\right) + B W_{m,\ell/2}\left(-\mathrm{i}\sqrt{C}\nu\right) \right\},$$

where

$$m = -\frac{\mathrm{i}\mu}{\sqrt{C}}.$$

In the case of $C = 0$, (5.7) is reduced to Bessel equation and its general solution can be represented as

$$G = A J_\ell \left(2\sqrt{-\mu}\sqrt{\nu} \right) + B H_\ell^{(1)} \left(2\sqrt{-\mu}\sqrt{\nu} \right).$$

Consider now (5.8) and define function $F(\eta)$. By introducing $b = g^2/f'$, we can rewrite (5.8) as

$$F' + \frac{f''}{2f'} F + \frac{b'}{2b} F = \frac{2\mathrm{i}\mu}{b} F,$$

where

$$F = \frac{1}{g} \exp \left(2\mathrm{i}\mu \int_{\eta_0}^{\eta} \frac{f' \mathrm{d}\eta}{g^2} \right).$$

In the last integral, one can assume that g depends on f, that is

$$g(\eta) = \gamma\left(f(\eta)\right), \tag{5.9}$$

which finally yields

$$F = \frac{1}{g} \exp \left(2\mathrm{i}\mu \int_{f_0}^{f} \frac{\mathrm{d}f}{\gamma^2} \right).$$

5.3 Surfaces Which Allow Variable Separation

As shown above, the parabolic equation written in coordinates (η, ν) allows variable separation only if the condition (5.6) is satisfied. This condition is the differential equation for function g, which defines the surface of the body as

$$r = k^{-2/3} g(\eta).$$

By using (5.9), this differential equation can be rewritten as

$$\gamma^3 \gamma'' = -C.$$

By noting that this equation does not contain independent variable f, one can reduce its order with the standard technique [9]. Assuming that $\gamma' \neq 0$ and considering the inverse function $f = H(\gamma)$, by introducing

$$D(\gamma) = \frac{d\gamma(f)}{df} = \frac{d\gamma\big(H(\gamma)\big)}{df}, \tag{5.10}$$

one gets

$$\dot{D} D = -\frac{C}{\gamma^3}.$$

Solving it, we find

$$D = \sqrt{D_0 + \frac{C}{\gamma^2}},$$

where D_0 is the constant of integration.

Now function γ can be found from (5.10)

$$f = \int \frac{d\gamma}{D(\gamma)} = \int \frac{d\gamma}{\sqrt{D_0 + C\gamma^{-2}}}.$$

If D_0 is not zero, we get

$$f = \frac{1}{\sqrt{D_0}} \int \frac{\gamma \, d\gamma}{\sqrt{C/D_0 + \gamma^2}} = \frac{1}{\sqrt{D_0}} \sqrt{\frac{C}{D_0} + \gamma^2},$$

and

$$\gamma(f) = \sqrt{D_0 f^2 - \frac{C}{D_0}}. \tag{5.11}$$

If $D_0 = 0$, then

$$f = \frac{1}{2\sqrt{C}}\gamma^2,$$

and

$$\gamma(f) = \sqrt{2}C^{1/4}\sqrt{f}. \tag{5.12}$$

We note also that in the particular case when $D_0 = 0$ and $C = 0$,

$$\gamma(f) = const.$$

Let us study the solutions $\gamma(f)$, derived above, and find out which types of surfaces allow variable separation in the corresponding parabolic equations. Evidently, the shape of the surface does not depend on function $f(\eta)$, which defines only the parameterization of the surface. Therefore, we further assume that $f(\eta) = \eta$. This makes $\gamma \equiv g$. As it was already noted, the case of $C = D_0 = 0$ corresponds to $g = const$ which is the case of a circular cylinder. Considering formula (5.11) in the case of C and D_0 both positive, one concludes that g is real only for $|\eta|$ larger than $\sqrt{C}D_0^{-2}$. The surface which is defined by $\nu = 1$ in that case is the two-sheeted hyperboloid. If $C < 0$ and $D_0 > 0$, then the surface is the one-sheeted hyperboloid. If $C > 0$ and $D_0 < 0$, then $|\eta|$ should be less than $\sqrt{C}D_0^{-2}$ and this is the case of the spheroid. The case when both C and D_0 are negative results in purely imaginary values of function g, so this case does not correspond to any surface. If $C = 0$, but $D_0 \neq 0$, then according to (5.11) the solution is

$$g = \sqrt{D_0}\eta,$$

which defines the conical surface. Finally, in the case of $D_0 = 0$ since the solution is given by formula (5.12), one needs to take positive C, and this is the case of the paraboloidal surface. The above observations are combined in Table 5.1, which lists all the canonical surfaces that yield parabolic equations allowing variable separation.

Table 5.1 Canonical surfaces

No.	C	D_0	$g(\eta)$	Surface
1	+	+	$g_1 = \sqrt{\eta^2 - 1}$	Two-sheeted hyperboloid
2	+	0	$g_2 = \sqrt{\eta}$	Paraboloid
3	+	–	$g_3 = \sqrt{1 - \eta^2}$	Spheroid
4	0	+	$g_4 = \eta$	Cone
5	0	0	$g_5 = const$	Cylinder
6	0	–	$g_6 = \mathrm{i}\eta$	
7	–	+	$g_7 = \sqrt{1 + \eta^2}$	One-sheeted hyperboloid
8	–	0	$g_8 = \mathrm{i}\sqrt{\eta}$	
9	–	–	$g_9 = \mathrm{i}\sqrt{1 + \eta^2}$	

It is easy to check that

$$\sqrt{|D_0|} = \frac{a}{b}, \qquad \sqrt{C} = \frac{a^2}{b} = \rho_e,$$

where a and b are the semiaxes of the body. In the case of spheroid D_0 is negative, and in the case of hyperboloid D_0 is positive; ρ_e is the radius of curvature at the end-points of the spheroid and two-sheeted hyperboloids. For cone it is equal to zero. For hyperboloids and cone, $\sqrt{D_0}$ is the tangent of the half angle between the asymptotes.

In the following sections of this chapter, we briefly consider the problems of diffraction by the new shapes, namely by the strongly elongated two-sheeted hyperboloid, by one-sheeted hyperboloid, by paraboloid and by cone. We do not consider the case of the cylinder, since it has no longitudinal scale and therefore there is no characteristic size of the surface which can be used to distinguish the case of the strongly elongated surface.

5.4 Strongly Elongated Two-Sheeted Hyperboloid

5.4.1 *Boundary-Layer Coordinates and the Parabolic Equation*

Consider the problem of diffraction of the plane acoustic wave by the right sheet of the two-sheeted hyperboloid. In this case, we define $g = g_1 = \sqrt{\eta^2 - 1}$ and consider $\eta > 1$. The surface of the hyperboloid is given by the equation $\nu = \nu_0$. To satisfy the requirement of the strongly elongated surface, we choose ν_0 to be of appropriate order in k. Coordinate η plays the role of one of the boundary-layer coordinates, while coordinate ν needs to be scaled. Instead of using the definition of coordinates according to formulas (5.1), in the case of the hyperboloid it is more convenient to use spheroidal coordinates. For that, we start with the definition of the hyperboloid as the surface given by the equation

$$\frac{z^2}{b^2} - \frac{x^2 + y^2}{a^2} = 1,$$

where a and b are the semiaxes, and introduce spheroidal coordinates (η, ξ, φ)

$$\begin{cases} x = p\sqrt{\eta^2 - 1}\sqrt{1 - \xi^2}\cos\varphi, \\ y = p\sqrt{\eta^2 - 1}\sqrt{1 - \xi^2}\sin\varphi, \\ z = p\eta\xi, \end{cases} \tag{5.13}$$

where $p = \sqrt{a^2 + b^2}$ is the half distance between the focuses. For consistency with the coordinates used in this chapter, we interchange in (5.13) the notations η and ξ as compared to Chaps. 2–4.

The surface of the hyperboloid coincides with the coordinate surface $\xi = \xi_0 = b/p$. If the hyperboloid is narrow (strongly elongated), then $a \ll b$ and $\xi_0 \approx 1$. In order to get the second boundary-layer coordinate (instead of ν), we stretch coordinate ξ as follows:

$$\xi = 1 - \frac{\chi\tau}{2kp}, \qquad \xi_0 = 1 - \frac{\chi}{2kp}.$$

Then, the surface is given by the equation $\tau = 1$.

Exploiting the symmetry of revolution, we represent the field in the form of Fourier series with respect to angle φ and denote the Fourier harmonics as u_n. The separation of the quick factor

$$u_n = \exp(\mathrm{i}kp\eta)U_n(\eta, \xi)$$

reduces the Helmholtz equation to the recurrent system of parabolic equations. At the leading order, we get

$$L_{n^2}U_n = 0, \tag{5.14}$$

where

$$L_s = \tau\frac{\partial^2}{\partial\tau^2} + \frac{\partial}{\partial\tau} + \frac{\mathrm{i}\chi}{2}\left(\eta^2 - 1\right)\frac{\partial}{\partial\eta} + \frac{\mathrm{i}\chi}{2}\eta + \frac{\chi^2\tau}{4} - \frac{s}{4\tau}.$$

In the case of Maxwell equations, we get the system of parabolic equations

$$\begin{cases} L_{1+n^2}E_n + \dfrac{in}{2\tau}H_n = 0, \\ L_{1+n^2}H_n - \dfrac{in}{2\tau}E_n = 0, \end{cases} \tag{5.15}$$

where E_n and H_n are the attenuation functions for the transverse components of the electric and magnetic fields. By introducing the new unknowns A and B as

$$E_n = A_n + B_n, \qquad H_n = -\mathrm{i}(A_n - B_n),$$

the system (5.15) is split into two independent parabolic equations

$$L_{(n-1)^2}A_n = 0, \qquad L_{(n+1)^2}B_n = 0.$$

An elementary solution of the parabolic equation (5.14) can be written as

$$\frac{1}{\sqrt{\eta^2 - 1}\sqrt{\nu}}\left(\frac{\eta + 1}{\eta - 1}\right)^{\mu} Y_{\mu,n/2}(-\mathrm{i}\chi\tau)$$

with μ being the parameter of variable separationand Y being a solution of the Whittaker equation, which can be represented as a combination of Whittaker functions M and W. If we integrate with respect to the parameter μ, we get the general representation for the attenuation function of the total field. Setting $\mu = it$ for consistency In the case of the spheroid, it can be written as

$$U_n = \frac{1}{\sqrt{\eta^2-1}\sqrt{\chi\tau}} \int \left(\frac{\eta+1}{\eta-1}\right)^{it} \left(a_n(t) M_{\mathrm{i}t,n/2}(-\mathrm{i}\nu) + b_n(t) W_{\mathrm{i}t,n/2}(-\mathrm{i}\chi\tau)\right) dt, \tag{5.16}$$

where $a_n(t)$ and $b_n(t)$ are arbitrary functions. As has been previously, this arbitrariness will be eliminated by matching with the incident wave and by requiring the boundary conditions to be satisfied.

5.4.2 Axial Incidence

Consider first the case of axial incidence. The nearest goal is to find the representation for the incident wave

$$U^{\mathrm{inc}} = \exp\left(-\frac{\mathrm{i}}{2}\eta\nu\right)$$

in the form of (5.16). Evidently, the only harmonics is with $n = 0$ in that case and all the other Fourier harmonics nullify. The problem consists of choosing the path of integration, a particular solution $Y_{\mathrm{i}t,0}(-\mathrm{i}\nu)$ of the Whittaker equation and determination of the amplitude multiplier such that

$$\exp\left(-\frac{\mathrm{i}}{2}\eta\nu\right) = \frac{1}{\sqrt{\eta^2-1}\sqrt{\nu}} \int \left(\frac{\eta+1}{\eta-1}\right)^{it} a_0(t) Y_{\mathrm{i}t,0}(-\mathrm{i}\nu) \mathrm{d}t. \tag{5.17}$$

If we consider this equality with fixed τ, then the integral transform, considered in (A.34), can be identified. By using the inversion, given by (A.35), we get for function $f(t)$ the following expression:

$$f(t) = \frac{\sqrt{\chi\tau}}{\pi} \int\limits_{(-\infty,-1]\cup[1,+\infty)} \mathrm{e}^{-\mathrm{i}\chi\tau\eta/2} \sqrt{\eta^2-1} \left(\frac{\eta+1}{\eta-1}\right)^{-\mathrm{i}t-1} \frac{\mathrm{d}\eta}{(\eta-1)^2}.$$

On the interval $\eta \in (-\infty, -1]$, we introduce a new variable σ as $\eta = -1 - 2\sigma$; on the interval $\eta \in [1, +\infty)$, the new variable is introduced such that $\eta = 1 + 2\sigma$. Then

$$f(t) = \frac{\sqrt{\chi\tau}}{\pi} e^{\mathrm{i}\chi\tau/2} \int_0^{+\infty} e^{\mathrm{i}\chi\tau\sigma} \sigma^{-\mathrm{i}t-1/2} (\sigma+1)^{\mathrm{i}t-1/2} \mathrm{d}\sigma$$

$$+ \frac{\sqrt{\chi\tau}}{\pi} e^{-\mathrm{i}\chi\tau/2} \int_0^{+\infty} e^{-\mathrm{i}\chi\tau\sigma} \sigma^{\mathrm{i}t-1/2} (\sigma+1)^{-\mathrm{i}t-1/2} \mathrm{d}\sigma. \tag{5.18}$$

Comparing the integrals in the right-hand side of (5.18) with the integral representation for the Whittaker function W, see (A.10), we conclude that

$$f(t) = \frac{e^{\mathrm{i}\pi/4}}{\pi} \Gamma\left(\frac{1}{2} - \mathrm{i}t\right) W_{\mathrm{i}t,0}(-\mathrm{i}\chi\tau) + \frac{e^{-\mathrm{i}\pi/4}}{\pi} \Gamma\left(\frac{1}{2} + \mathrm{i}t\right) W_{-\mathrm{i}t,0}(\mathrm{i}\chi\tau). \tag{5.19}$$

It is worth noting that both Whittaker functions $W_{\mathrm{i}t,0}(-\mathrm{i}\chi\tau)$ and $W_{-\mathrm{i}t,0}(\mathrm{i}\chi\tau)$ satisfy the same differential equation. It is worth noting that, when considering the integral on the right-hand side of (5.17), one can exploit the analyticity of the integrated functions. The first term in (5.19) contains no singularities in the upper half-plane of complex variable t. Since the ratio $(\eta+1)/(\eta-1)$ is greater than unity in the considered domain of $\eta > 1$, the path of integration can be closed by adding the arc of infinite radius in $\mathrm{Im}\, t > 0$ and the integral will be equal to zero. Therefore, one can take as well the solution

$$f(t) = \frac{e^{-\mathrm{i}\pi/4}}{\pi} \Gamma\left(\frac{1}{2} + \mathrm{i}t\right) W_{-\mathrm{i}t,0}(\mathrm{i}\chi\tau),$$

which gives the identity

$$\exp\left(-\frac{\mathrm{i}}{2}\eta\chi\tau\right) = \frac{e^{-\mathrm{i}\pi/4}}{\pi\sqrt{\eta^2-1}\sqrt{\chi\tau}} \int_{-\infty}^{+\infty} \left(\frac{\eta+1}{\eta-1}\right)^{\mathrm{i}t} \Gamma\left(\frac{1}{2} + \mathrm{i}t\right) W_{-\mathrm{i}t,0}(\mathrm{i}\chi\tau) \mathrm{d}\lambda. \tag{5.20}$$

It is worth reminding that $\eta > 1$ and $\chi\tau > 0$ in this formula. To achieve convergence of the integral on the right-hand side of (5.20), one shifts the path of integration for positive t to the upper half-plane.

Moreover, one can express the Whittaker function $W_{-\mathrm{i}t,0}(\mathrm{i}\chi\tau)$ via functions $M_{\mathrm{i}t,0}(-\mathrm{i}\chi\tau)$ and $W_{\mathrm{i}t,0}(-\mathrm{i}\chi\tau)$; see (A.13). Then the right-hand side of identity (5.20) can be rewritten as

$$\frac{e^{-\mathrm{i}\pi/4}}{\sqrt{\eta^2-1}\sqrt{\chi\tau}} \int_{-\infty}^{+\infty} \left(\frac{\eta+1}{\eta-1}\right)^{\mathrm{i}t} \frac{e^{\pi\lambda}}{\cosh(\pi t)} M_{\mathrm{i}t,0}(-\mathrm{i}\chi\tau) \mathrm{d}t$$

$$+ \frac{e^{\mathrm{i}\pi/4}}{\pi\sqrt{\eta^2-1}\sqrt{\chi\tau}} \int_{-\infty}^{+\infty} \left(\frac{\eta+1}{\eta-1}\right)^{\mathrm{i}t} \Gamma\left(\tfrac{1}{2} - \mathrm{i}t\right) W_{\mathrm{i}t,0}(-\mathrm{i}\chi\tau) \mathrm{d}t. \tag{5.21}$$

Here, we have taken into account the symmetry formula [1] for the gamma-function

$$\Gamma\left(\frac{1}{2}+\mathrm{i}t\right)\Gamma\left(\frac{1}{2}-\mathrm{i}t\right)=\frac{\pi}{\cosh(\pi t)}.$$

The second integral in (5.21) is again equal to zero if $\eta > 1$, thus we get another identity

$$\frac{E^{-\mathrm{i}\pi/4}}{\sqrt{\eta^2-1}\sqrt{\chi\tau}}\int\limits_{-\infty}^{+\infty}\left(\frac{\eta+1}{\eta-1}\right)^{\mathrm{i}t}\frac{\mathrm{e}^{\pi t}}{\cosh(\pi t)}M_{\mathrm{i}t,0}(-\mathrm{i}\chi\tau)\mathrm{d}t=\exp\left(-\frac{\mathrm{i}}{2}\eta\chi\tau\right), \quad (5.22)$$

valid for $\eta > 1$ and $\chi\tau > 0$.

It is worth comparing (5.22) with the analytic continuation of the representation (A.40), which expresses the incident plane wave in the coordinates of the boundary layer on the surface of strongly elongated spheroid. For the case of axial incidence, it can be simplified to

$$\frac{\mathrm{e}^{\mathrm{i}\pi/4}}{\sqrt{1-\eta^2}\sqrt{\chi\tau}}\int\limits_{-\infty}^{+\infty}\left(\frac{1-\eta}{1+\eta}\right)^{\mathrm{i}t}M_{\mathrm{i}t,0}(-\mathrm{i}\chi\tau)\frac{\mathrm{d}t}{\cosh(\pi t)}=\exp\left(\frac{\mathrm{i}}{2}\eta\chi\tau\right). \quad (5.23)$$

By considering $\eta > 1$ in (5.23) and performing the transformations

$$\sqrt{1-\eta^2}=\mathrm{i}\sqrt{\eta^2-1}, \qquad \frac{1-\eta}{1+\eta}=-\frac{\eta-1}{\eta+1}, \qquad (-1)^{\mathrm{i}\lambda}=e^{\mathrm{i}\pi\mathrm{i}t}=\mathrm{e}^{-\pi t},$$

after replacing t with $-t$ on the left-hand side, we get

$$\frac{\mathrm{e}^{-\mathrm{i}\pi/4}}{\sqrt{\eta^2-1}\sqrt{\chi\tau}}\int\limits_{-\infty}^{+\infty}\left(\frac{\eta+1}{\eta-1}\right)^{\mathrm{i}t}\frac{\mathrm{e}^{\pi t}}{\cosh(\pi t)}M_{-\mathrm{i}t,0}(-\mathrm{i}\chi\tau)\mathrm{d}t.$$

Then basing on (A.11), one can substitute $\mathrm{i}M_{\mathrm{i}t,0}(\mathrm{i}\chi\tau)$ instead of $M_{-\mathrm{i}t,0}(-\mathrm{i}\chi\tau)$ and change the sign of τ. This transforms (5.23) to (5.22).

After discussing the representations for the incident field, we turn to finding the diffracted field. Whittaker function $W_{-\mathrm{i}t,0}(\mathrm{i}\chi\tau)$, presented in the representation (5.20), corresponds only to the incoming waves, which can be seen from its asymptotic expansion (A.23). In view of the radiation conditions, one should choose the Whittaker function $W_{\mathrm{i}t,0}(-\mathrm{i}\chi\tau)$ in the representation for the reflected waves. Thus, the solution to the diffraction problem for plane wave incidence on the right sheet of the two-sheeted hyperboloid requires finding the amplitude multiplier at this Whittaker function. For this, we use the boundary conditions. As the result, the leading order approximation for the total field can be written as

$$u = \exp(\mathrm{i}kp\eta)\,\frac{\mathrm{e}^{-\mathrm{i}\pi/4}}{\pi\sqrt{\chi\tau}\sqrt{\eta^2-1}}\int\limits_{-\infty}^{+\infty}\left(\frac{\eta+1}{\eta-1}\right)^{it}\Gamma\left(\frac{1}{2}+\mathrm{i}t\right)\times$$
$$\times\left\{W_{-\mathrm{i}t,0}(\mathrm{i}\chi\tau)+R(t)W_{\mathrm{i}t,0}(-\mathrm{i}\chi\tau)\right\}\mathrm{d}t.$$

In the case of the Dirichlet boundary condition, the reflection coefficient $R(t)$ has the form

$$R = R_D = -\frac{W_{-\mathrm{i}t,0}(\mathrm{i}\chi)}{W_{\mathrm{i}t,0}(-\mathrm{i}\chi)},$$

and for acoustically hard surface, it is

$$R = R_N = \frac{2\chi\dot{W}_{-\mathrm{i}t,0}(\mathrm{i}\chi)+\mathrm{i}W_{-\mathrm{i}t,0}(\mathrm{i}\chi)}{2\chi\dot{W}_{\mathrm{i}t,0}(-\mathrm{i}\chi)-\mathrm{i}W_{\mathrm{i}t,0}(-\mathrm{i}\chi)}.$$

In the formulas for the field and its normal derivative on the surface of the hyperboloid, we take into account the expression for the Wronskian of the Whittaker functions

$$\dot{W}_{\mathrm{i}t,0}(-\mathrm{i}\nu)W_{-\mathrm{i}t,0}(\mathrm{i}\nu)+W_{\mathrm{i}t,0}(-\mathrm{i}\nu)\dot{W}_{-\mathrm{i}t,0}(\mathrm{i}\nu) = -\exp(\pi t).$$

The resulting expression for the case of the Dirichlet problem is

$$\left.\frac{\partial u}{\partial\tau}\right|_{\partial\Omega} = \exp(\mathrm{i}kp\eta)\,\frac{\mathrm{e}^{3\mathrm{i}\pi/4}\sqrt{\chi}}{\pi\sqrt{\eta^2-1}}\int\limits_{-\infty}^{+\infty}\left(\frac{\eta+1}{\eta-1}\right)^{it}\frac{\Gamma(1/2+\mathrm{i}t)\mathrm{e}^{\pi t}}{W_{\mathrm{i}t,0}(-\mathrm{i}\chi)}\mathrm{d}t. \quad (5.24)$$

In the case of an absolutely rigid surface, i.e. Neumann problem, the field on the surface has the asymptotics

$$u|_{\partial\Omega} = \exp(\mathrm{i}kp\eta)\frac{\mathrm{e}^{3\mathrm{i}\pi/4}}{\pi\sqrt{\chi}\sqrt{\eta^2-1}}\times$$
$$\times\int\limits_{-\infty}^{+\infty}\left(\frac{\eta+1}{\eta-1}\right)^{it}\frac{\Gamma(1/2+\mathrm{i}t)\mathrm{e}^{\pi t}}{\dot{W}_{\mathrm{i}t,0}(-\mathrm{i}\chi)-\frac{\mathrm{i}}{2\chi}W_{\mathrm{i}t,0}(-\mathrm{i}\chi)}\mathrm{d}t. \quad (5.25)$$

The expressions under the sign of the integrals in (5.24) and (5.25) rapidly decrease on the negative semiaxis, but the convergence of the integrals on the positive semiaxis is only due to oscillations of the multiplier

$$\left(\frac{\eta+1}{\eta-1}\right)^{it}.$$

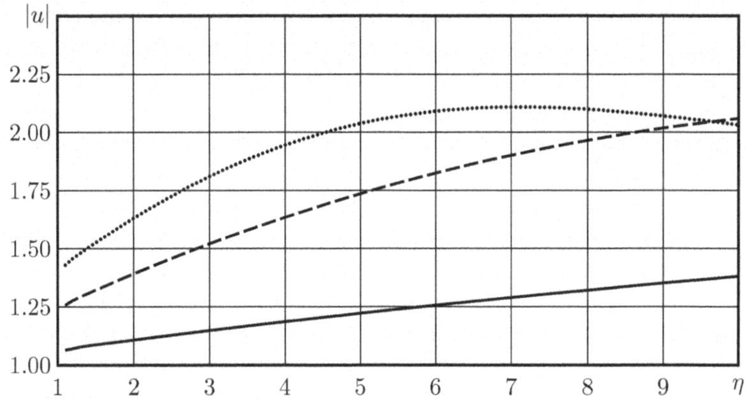

Fig. 5.1 Fields on the surface of absolutely rigid hyperboloids with $\chi = 0.1$ (solid line), $\chi = 0.5$ (dashed line) and $\chi = 1$ (dotted line)

Since fraction $(\eta + 1)/(\eta - 1) > 1$ for the considered values of η, the path of integration can be shifted from the positive semiaxis to the upper half-plane, which improves the convergence of the integral.

The distributions of the field intensities on the surfaces of narrow hyperboloids are presented in Fig. 5.1.

5.4.3 Skew Incidence

The case of skew incidence can be studied by the method of analytic continuation. For that, we use the approximations valid in the case of diffraction by the strongly elongated spheroid and do the same replacements as when transforming from (5.23) to (5.22). The result, which can be checked by direct substitution in the parabolic equation and the boundary conditions, is as follows:

$$U_n = \frac{(2 - \delta_0^n)\mathrm{i}^n}{\pi\sqrt{\eta^2 - 1}\sqrt{\chi\tau}\alpha} \int_{-\infty}^{\mathrm{e}^{\mathrm{i}\pi/4}\infty} \left(\frac{\eta + 1}{\eta - 1}\right)^{\mathrm{i}t} M_{\mathrm{i}t,n/2}(-\mathrm{i}\alpha^2) \times$$
$$\times \left\{ M_{\mathrm{i}t,n/2}(-\mathrm{i}\chi\tau) + R_n(t) W_{\mathrm{i}t,n/2}(-\mathrm{i}\chi\tau) \right\} \Omega_n(t)\mathrm{e}^{\pi t}\mathrm{d}t, \qquad (5.26)$$

where $\alpha = \sqrt{kp}\theta_0$ is the scaled incidence angle and functions $\Omega_n(t)$ and $R_n(t)$ are given by the same formulas (4.30) and (3.19), (3.20) as in the case of diffraction by the strongly elongated spheroid.

Formulas presented here are obtained by means of analytic continuation. However, the same expressions can be derived from the usual technique of the parabolic equation method adopted to strongly elongated bodies [3].

It is worth noting that computations according to formula (5.26) face some difficulties caused by the necessity to shift the path of integration to the upper complex half-plane of variable t in order to achieve better convergence of the integral. When transforming to the use of Coulomb wave functions, which are effectively evaluated by the code developed in [15], one comes to the necessity to compute Coulomb wave functions with the first argument being complex. However, it was found in [3] that the upper half-plane of parameter t contains a region where the COULCC code fails to compute the values of the Coulomb wave functions accurately, and the integration path goes through this region. In the case of small values of η, this, nevertheless, causes no harm, but when η increases, the results of integration become unstable. To avoid this problem, one can exploit the recurrence rule

$$W_{\mu,n}(z) = \left(z + 2 + 2\mu\right) W_{\mu-1,n}(z) - \left(\left(\frac{3}{2} - \mu\right)^2 - n^2\right) W_{\mu-2,n}(z),$$

which follows from formulas 9.234 1 and 2 of [8]. With the help of this rule, applied several times, one can avoid the necessity to compute Whittaker functions with the indices being in the region of unstable computations.

5.5 Strongly Elongated Paraboloid

Consider now the problem of diffraction by the narrow paraboloid, which corresponds to $g = g_2 = \sqrt{\eta}$. Let its surface be defined by the equation $w = \chi$ in the paraboloidal coordinates (v, w, φ) introduced by the formulas (compared to (4.4))

$$x = \frac{\sqrt{vw}}{k} \sin\varphi, \quad y = \frac{\sqrt{vw}}{k} \cos\varphi, \quad z = \frac{v - w}{2k}.$$

The axially incident plane wave e^{ikz}, when written in coordinates (v, w, φ), is represented as the quick factor $e^{iv/2}$ and the attenuation function

$$U^{(i)} = \exp\left(-\frac{i}{2} w\right).$$

That is, the variables are initially separated and the integral representation for the incident wave is reduced to formula

$$\exp\left(\frac{i}{2} w\right) = \frac{e^{i\pi/4}}{\sqrt{w}} M_{1/2,0}(-iw).$$

Then the representation for the total field can be written as

$$u = 2\exp\left(\mathrm{i}k^{2/3}\eta\right)\left\{\exp\left(-\frac{\mathrm{i}}{2}w\right) + R\exp\left(\frac{\mathrm{i}}{2}w\right)\right\},$$

where

$$R_D = -\exp\left(-\mathrm{i}\chi\right), \qquad R_N = \exp\left(-\mathrm{i}\chi\right).$$

These formulas do not add anything new to the problem of diffraction by the paraboloid, because in the problems of diffraction by a paraboloid the variables can be separated both in Helmholtz and Maxwell equations.

5.6 One-Sheeted Hyperboloid

Consider now the case of diffraction by the one-sheeted hyperboloid. The surface corresponds to $g = g_7 = \sqrt{1+\eta^2}$, and we can consider any (real) values of coordinate η. The spheroidal coordinates are now introduced by the formulas [10]

$$x = p\sqrt{1+\eta^2}\sqrt{1-\xi^2}\sin\varphi, \quad y = p\sqrt{1+\eta^2}\sqrt{1-\xi^2}\cos\varphi, \quad z = p\eta\xi,$$

where $p = \sqrt{b^2 - a^2}$. The surface is given by the equation $\xi = \xi_0$ and the requirement of the strongly elongated body asks ξ_0 to be close to unity. Introducing the stretched coordinate τ such that

$$\xi = 1 - \frac{\chi\tau}{2kp},$$

where

$$\chi = 2kp\Big(1 - \xi_0\Big),$$

we get the boundary-layer coordinates (η, τ, φ) with $\tau = 1$ on the surface.

By extracting the quick oscillating factor in the form

$$u = \exp(\mathrm{i}kp\eta)U(\eta, \tau, \varphi),$$

the Helmholtz equation reduces at the leading order of kp to the parabolic equation with the operator

$$L = \tau\frac{\partial^2}{\partial\tau^2} + \frac{\partial}{\partial\tau} + \frac{\mathrm{i}\chi}{2}(1+\eta^2)\frac{\partial}{\partial\eta} + \left(\frac{\mathrm{i}\chi}{2}\eta - \frac{\chi^2}{2}\tau - \frac{n^2}{4\tau}\right).$$

This operator differs from that in the case of spheroid by the replacements

$$\eta \to \mathrm{i}\eta, \qquad \tau \to \mathrm{i}\tau. \tag{5.27}$$

The same replacements applied to formula (3.17) of Chap. 3 yield the representation for the field in the boundary layer near the surface. It takes the form

$$U_n = \frac{1}{\sqrt{\chi\tau}\sqrt{1+\eta^2}} \int_{-\infty}^{+\infty} \exp\Big(2t\tan^{-1}(\eta)\Big) A_n(t) \times$$
$$\times \Big(M_{\mathrm{i}t,n/2}(\chi\tau) + R_n(t) W_{\mathrm{i}t,n/2}(\chi\tau)\Big)\mathrm{d}t.$$

For the incident wave representation, one can also apply the replacements (5.27) to formula (3.29), which is valid in the case of axially incident wave. This yields

$$\exp\left(-\frac{\mathrm{i}}{2}\eta\chi\tau\right) = \frac{1}{\sqrt{\chi\tau}\sqrt{1+\eta^2}} \int_{-\infty}^{+\infty} \exp\Big(2t\tan^{-1}(\eta)\Big) M_{\mathrm{i}t,0}(\chi\tau) \frac{\mathrm{d}t}{\cosh(\pi t)}.$$

Further, it is necessary to find the representation for the reflected wave. However, in the case of diffraction by the one-sheeted hyperboloid as well as in the case of diffraction by any concave boundary, it is difficult to separate the reflected wave from the incident one because any wave, outgoing from one part of the boundary, returns to the surface and becomes an incoming wave to the other part of the boundary. In the case of strongly elongated bodies, the distance between these two parts of the surface is large, and therefore it is hardly possible to invent the approximation which would correctly take into account the phase accumulated by the wave while traveling from one part of the surface to another. This circumstance appears to be the crucial obstacle that prevents us from finding the representation for the reflected wave and as a consequence, we cannot construct the asymptotic approximation for the diffracted field on the strongly elongated one-sheeted hyperboloid.

5.7 Narrow Cone

5.7.1 Problem of Diffraction

The case of $g = g_4 = \eta$ corresponds to the conical surface. It can be considered as the limiting case of hyperboloid, however, we do not exploit this limit. Instead, we start by specifying the cone as the surface given by the equation

$$\sqrt{x^2 + y^2} = \tan(\gamma) z. \tag{5.28}$$

The asymptotic representations for the problems of diffraction by the strongly elongated spheroid (see Chaps. 3 and 4) and by strongly elongated one-sheeted hyperboloid, considered in this chapter, are valid in the "middle" part of the surface. That is at some distance from the tips. In the same way, when considering the problem of diffraction by the cone, we construct the field representations in the domain being at some distance from the tip of the cone. More precisely, we consider such values of r and z coordinates that both kr and kz are large. Moreover, we suppose that

$$(kr)^2 \sim kz.$$

This can be rewritten as the requirement for the angle of the cone to be small

$$\gamma = \tan^{-1}\left(\frac{kr}{kz}\right) = O\left((kp)^{-1/2}\right),$$

where we fix some value $p = z$, where z is taken in the considered domain, and use kp as the large asymptotic parameter. Then we can define the parameter of elongation χ by the usual formula

$$\chi = \frac{kr^2}{kz} = kp\gamma^2$$

and assume it to be of order $O(1)$.

It is worth noting that problems of diffraction by cones are subjects of many researches; see, for example, [6, 7, 14]. The procedure of constructing the solution is always based on the separation of the radial coordinate in the spherical coordinate system with the center at the tip of the cone and the usual goal is in finding the representation for the wave that spreads away from the tip. This goal is explained by saying that the other part of the solution can be constructed by means of the ray method. However, if the cone is very narrow, the domain where the ray asymptotic representations may be used appears to be at very large distances from the tip, where the radius of curvature becomes much larger than the wavelength. This results in the appearance of the domain where local representations valid near the tip are already inapplicable, since the distance from the tip is large, but the ray approximation is also yet not applicable, since the radius of transverse curvature $\rho_t \approx p\gamma$ is not large enough. We construct the asymptotic approximation basing on the asymptotic smallness of the cone angle and intend to use it for the approximation of the field only in a domain being at some distance from the tip and near the surface.

In the boundary layer near the surface, we introduce a coordinate system (η, ν, φ), related to cylindrical coordinates (z, r, φ) by the formulas

$$\begin{cases} z = p\eta\left(1 - \dfrac{\nu}{2kL}\right), \\ x = p\eta\sqrt{\dfrac{\nu}{kp}}\sin\varphi, \\ y = p\eta\sqrt{\dfrac{\nu}{kp}}\cos\varphi. \end{cases} \tag{5.29}$$

Note that the second term in the formula for z is introduced to make the coordinate system (η, ν, φ) orthogonal at the leading order by kp. Generally speaking, it is not necessary to require the system of coordinates to be orthogonal and we do that only in order to simplify the intermediate derivations.

The surface of the cone in coordinates (η, ν, φ) is given by the equation

$$\nu = \chi.$$

We search for the solution u in the usual form for the parabolic equation method as the product of the quickly oscillating factor and the attenuation function, which we represent as the sum of Fourier harmonics with respect to the angle φ, i.e.

$$u = e^{ikp\eta} \sum_{n=-\infty}^{+\infty} U_n(\eta, \nu) e^{in\varphi}. \tag{5.30}$$

Substituting this representation into the Helmholtz equation yields at the leading order the parabolic equation

$$\nu \frac{\partial^2 U_n}{\partial \nu^2} + \frac{\partial U_n}{\partial \nu} + \frac{i}{2}\eta^2 \frac{\partial U_n}{\partial \eta} + \left(\frac{i}{2}\eta - \frac{n^2}{4\nu}\right) U_n = 0. \tag{5.31}$$

Thus, in the case of acoustic wave diffraction, the leading order approximation for the attenuation function is the solution to the boundary-value problem for the parabolic equation (5.31) with the boundary condition specified at $\nu = \chi$ and the condition at infinity, which matches the field with the representation of the ray method. We consider this problem in the following section, and the case of electromagnetic wave diffraction is considered in Sect. 5.7.3.

5.7.2 *Diffraction of the Plane Acoustic Wave*

We consider the two cases of the boundary conditions, namely acoustically hard cone with the Neumann boundary condition on its surface and acoustically soft cone with the Dirichlet boundary condition.

Let the incident plane wave run along the axis of the cone

$$u^{\text{inc}} = \mathrm{e}^{ikz}.$$

In this simpler case, the only nonzero Fourier harmonics in (5.30) is U_0.

Parabolic equation (5.31) allows variable separation whichresults in the representation of its solution in the form of the integral with respect to the parameter of variable separation

$$u = \mathrm{e}^{ikp\eta}\frac{1}{\eta}\int \exp\left(-\frac{\lambda}{\eta}\right) Z_0\left(\mathrm{e}^{\mathrm{i}\pi/4}\sqrt{2\lambda\nu}\right) a(\lambda)\mathrm{d}\lambda. \tag{5.32}$$

Here, we have written explicitly the solution of the first-order ordinary differential equation by variable η, and expressed the solution of the differential equation by variable ν via cylindrical function Z_0.

Both the total field and its parts, the incident plane wave u^{inc} and the wave u^{refl}, reflected from the surface of the cone, can be represented in the form (5.32). Evidently, the incident wave can be prolongated behind the surface of the cone, and it is regular there including the axis of the cone. This defines the choice of the Bessel function $Z_0 = J_0$ as the solution by variable ν for the incident wave. For the reflected wave in view of the radiation conditions, one should choose the Hankel function of the first kind, that is $Z_0 = H_0^{(1)}$. So,

$$u^{\text{inc}} = \mathrm{e}^{ikp\eta}\frac{1}{\eta}\int \exp\left(-\frac{\lambda}{\eta}\right) J_0\left(\mathrm{e}^{\mathrm{i}\pi/4}\sqrt{2\lambda\nu}\right) a(\lambda)\mathrm{d}\lambda, \tag{5.33}$$

$$u = \mathrm{e}^{ikp\eta}\frac{1}{\eta}\int \exp\left(-\frac{\lambda}{\eta}\right) a(\lambda)\Big\{J_0\left(\mathrm{e}^{\mathrm{i}\pi/4}\sqrt{2\lambda\nu}\right) + R(\lambda)H_0^{(1)}\left(\mathrm{e}^{\mathrm{i}\pi/4}\sqrt{2\lambda\nu}\right)\Big\}\mathrm{d}\lambda. \tag{5.34}$$

Here, functions $a(\lambda)$ and $R(\lambda)$ are to be determined; the second has the meaning of the reflection coefficient of an elementary wave.

Find at first function $a(\lambda)$ in the representation (5.33) Of the incident wave. For that, we use formula 6.614.1 from [8]

$$\int_0^{+\infty} \mathrm{e}^{-\alpha x} J_m\left(\beta\sqrt{x}\right) \mathrm{d}x = \frac{\beta}{4}\sqrt{\frac{\pi}{\alpha^3}} \exp\left(-\frac{\beta^2}{8\alpha}\right) \times \left[I_{\frac{m-1}{2}}\left(\frac{\beta^2}{8\alpha}\right) - I_{\frac{m+1}{2}}\left(\frac{\beta^2}{8\alpha}\right)\right]. \tag{5.35}$$

Since the modified Bessel functions of the semi-integer index are expressed via the elementary functions

$$I_{1/2}(z) = \sqrt{\frac{2z}{\pi}}\frac{\sinh(z)}{z}, \qquad I_{-1/2}(z) = \sqrt{\frac{2z}{\pi}}\frac{\cosh(z)}{z},$$

the right-hand side of formula (5.35) for $m = 0$ simplifies. We have

$$\int\limits_0^{+\infty} \mathrm{e}^{-\alpha x} J_0\left(\beta\sqrt{x}\right) \mathrm{d}x = \frac{1}{\alpha}\exp\left(-\frac{\beta^2}{4\alpha}\right). \tag{5.36}$$

The attenuation function corresponding to the incident wave is

$$U^{\text{inc}} = \exp\left(-\frac{\mathrm{i}}{2}\eta\nu\right).$$

Comparing this expression with the right-hand side of formula (5.36), we set $\alpha = \eta^{-1}$, $\beta = \mathrm{e}^{\mathrm{i}\pi/4}\sqrt{2\nu}$, wherefrom we get for the incident wave the required representation (5.33) with

$$a(\lambda) = 1.$$

Now let us find function $R(\lambda)$ in formula (5.34) for the total field. It should be taken such that the boundary conditions are satisfied on the surface of the cone. In the case of the Neumann boundary condition, we get

$$R = R_N = -\frac{J_1\left(\mathrm{e}^{\mathrm{i}\pi/4}\sqrt{2\lambda\chi}\right)}{H_1^{(1)}\left(\mathrm{e}^{\mathrm{i}\pi/4}\sqrt{2\lambda\chi}\right)},$$

and in the case of the Dirichlet boundary condition

$$R = R_D = -\frac{J_0\left(E^{\mathrm{i}\pi/4}\sqrt{2\lambda\chi}\right)}{H_0^{(1)}\left(\mathrm{e}^{\mathrm{i}\pi/4}\sqrt{2\lambda\chi}\right)}.$$

Substituting these functions in the representation (5.34) for the total field, one can find the total field on the surface of the cone. When simplifying the formulas, we take into account the expression for the Wronskian

$$J_0(x)\dot{H}_0^{(1)}(x) - \dot{J}_0(x)H_0^{(1)}(x) = \frac{2\mathrm{i}}{\pi x}.$$

In the Neumann problem, we get the following asymptotics of the field on the surface:

$$u\Big|_{\nu=\chi} \sim \mathrm{e}^{\mathrm{i}kp\eta} F_N\left(2kp\gamma^2\eta\right). \tag{5.37}$$

Here, we expressed the value χ via the cone angle γ and introduced the special function

$$F_N(\xi) = \frac{4e^{5i\pi/4}}{\pi\xi} \int_0^{+\infty} \exp\left(-\frac{t^2}{\xi}\right) \frac{dt}{H_1^{(1)}\left(e^{i\pi/4}t\right)}.$$

In the case of the Dirichlet problem, we compute the normal derivative of the total field on the surface. In view of

$$\frac{\partial}{\partial n} \approx \frac{\partial}{\partial r} = \frac{2\sqrt{kp\chi}}{p\eta} \frac{\partial}{\partial \nu},$$

we find

$$\frac{\partial u}{\partial n} \sim k\gamma e^{ikp\eta} F_D\left(2kp\gamma^2\eta\right), \tag{5.38}$$

where

$$F_D(\xi) = -\frac{8i}{\pi\xi^2} \int_0^{+\infty} \exp\left(-\frac{t^2}{\xi}\right) \frac{t\,dt}{H_0^{(1)}\left(e^{i\pi/4}t\right)}.$$

Special functions $F_N(\xi)$ and $F_D(\xi)$ are presented in Fig. 5.2. For the computations, the upper limit of integration was replaced with the value $t = 100$.

Note that expressions (5.37) and (5.38) first appeared in [4] and very similar expressions have been derived in [12] by using a completely different technique. The case of the incidence at an angle to the axis is considered in [2]. The derivations are similar, and we present here only the final result

$$U_n = \frac{1}{\eta} \int_0^{+\infty} \exp\left(-\frac{\lambda}{\eta}\right) \Omega_n(\lambda) \times$$
$$\times \left\{ J_n\left(e^{i\pi/4}\sqrt{2\lambda\nu}\right) + R_n(\lambda) H_n^{(1)}\left(e^{i\pi/4}\sqrt{2\lambda\nu}\right) \right\} d\lambda,$$

where Ω_n is defined by the incident wave and depends on the angle of incidence ϑ_0:

$$\Omega_n(\lambda) = J_{|n|}\left(e^{i\pi/4}\sqrt{2kp\lambda}\vartheta_0\right)$$

and coefficients R_n depend on whether we consider the Dirichlet or the Neumann boundary condition:

$$R_n^D = -\frac{J_{|n|}\left(e^{i\pi/4}\sqrt{2\lambda\chi}\right)}{H_{|n|}^{(1)}\left(e^{i\pi/4}\sqrt{2\lambda\chi}\right)}, \qquad R_n^N = -\frac{\dot{J}_{|n|}\left(e^{i\pi/4}\sqrt{2\lambda\chi}\right)}{\dot{H}_{|n|}^{(1)}\left(e^{i\pi/4}\sqrt{2\lambda\chi}\right)}.$$

We remark that when deriving the above formulas the angle of incidence ϑ_0 was assumed to be small, so $\sqrt{kp}\vartheta_0 < const = O(1)$.

5.7.3 Diffraction of the Plane Electromagnetic Wave

Consider now the problem of diffraction of the plane electromagnetic wave by a narrow perfectly conducting circular cone. We consider only the case of axial incidence, though the incidence at a small angle to the axis can be considered similarly. In a cylindrical coordinate system, the dependence of the electromagnetic field components on the angle φ is given by the multipliers $\cos\varphi$ or $\sin\varphi$. We separate all these multipliers and with the help of four scalar equations from the system of Maxwell equations express the components of the field via H_φ and E_φ. We remind that as in Chap. 4, we work with Maxwell equation rewritten in symmetric form

$$\begin{cases} \nabla \times \mathbf{E} = \mathrm{i}k\mathbf{H}, \\ \nabla \times \mathbf{H} = -\mathrm{i}k\mathbf{E}, \end{cases}$$

for vector functions $\mathbf{E}$ expressing the electric field divided by the impedance of the space and $\mathbf{H}$, which is the magnetic field. For the components H_φ and E_φ, the remaining Maxwell equations yield the system of the second-order differential equations which can be easily reduced to two independent equations

$$L_{(-)}A = 0, \qquad L_{(+)}B = 0,$$

where

$$L_{(\pm)} = \frac{\partial^2}{\partial z^2} \pm \frac{2\mathrm{i}k}{(kr)^2 - 1}\frac{\partial}{\partial z} + \frac{\partial^2}{\partial r^2} + \frac{1}{r}\left(1 - \frac{2}{(kr)^2 - 1}\right)\frac{\partial}{\partial r} + \left(\frac{(kr)^2 - 1}{r^2} - \frac{1}{r^2}\frac{(kr)^2 + 1}{(kr)^2 - 1}\right),$$

for functions

$$A = \frac{1}{2}H_\varphi - \frac{\mathrm{i}}{2}E_\varphi, \qquad B = \frac{1}{2}H_\varphi + \frac{\mathrm{i}}{2}E_\varphi. \tag{5.39}$$

We rewrite these equations in coordinates (η, ν) introduced by formulas (5.29) and similar to (5.30) extract the quickly oscillating multiplier

$$A = \mathrm{e}^{\mathrm{i}kp\eta}\Phi(\eta, \nu), \qquad B = \mathrm{e}^{\mathrm{i}kp\eta}\Psi(\eta, \nu).$$

At the leading order by kp, we get the parabolic equations

$$L_0\Phi = 0, \qquad L_1\Psi = 0,$$

where

$$L_n = \nu\frac{\partial^2}{\partial \nu^2} + \frac{\partial}{\partial \nu} + \frac{\mathrm{i}}{2}\eta^2\frac{\partial}{\partial \eta} + \frac{\mathrm{i}}{2}\eta - \frac{n^2}{\nu}.$$

Similar to the scalar case, these equations allow variable separation,and as the result we get the representations for the solutions

$$\Phi = \frac{1}{\eta} \int \exp\left(-\frac{\lambda}{\eta}\right) \phi(\lambda) Z_0 \left(\mathrm{e}^{\mathrm{i}\pi/4}\sqrt{2\lambda\nu}\right) \mathrm{d}\lambda,$$

$$\Psi = \frac{1}{\eta} \int \exp\left(-\frac{\lambda}{\eta}\right) \psi(\lambda) Z_2 \left(E^{\mathrm{i}\pi/4}\sqrt{2\lambda\nu}\right) D\lambda,$$

where Z_j are the cylindrical functions. In the representation for the incident field, we choose Bessel function as in the scalar case and for the reflected field we choose the Hankel function of the first kind which agrees with the radiation principle.

First, we obtain the representation for the incident plane wave

$$\mathbf{E}^{\mathrm{inc}} = \mathrm{e}^{ikz}\mathbf{e}_x, \qquad \mathbf{H}^{\mathrm{inc}} = \mathrm{e}^{ikz}\mathbf{e}_y.$$

After the transformation to functions (5.39) and separation of the quick oscillating factor, we get

$$\Phi^{\mathrm{inc}} = \frac{1}{2}\exp\left(-\frac{\mathrm{i}}{2}\eta\nu\right), \qquad \Psi^{\mathrm{inc}} = 0.$$

Using the above-derived representation for the incident acoustic wave and switching back to the initial components of the electromagnetic field, we get

$$E_\varphi^{\mathrm{inc}} = \mathrm{e}^{ikp\eta}\frac{\sin\varphi}{\eta} \int\limits_0^{+\infty} \mathrm{e}^{-\lambda/\eta} J_0 \left(\mathrm{e}^{\mathrm{i}\pi/4}\sqrt{2\lambda\nu}\right) \mathrm{d}\lambda,$$

$$H_\varphi^{\mathrm{inc}} = \mathrm{e}^{ikp\eta}\frac{\cos\varphi}{\eta} \int\limits_0^{+\infty} \mathrm{e}^{-\lambda/\eta} J_0 \left(\mathrm{e}^{\mathrm{i}\pi/4}\sqrt{2\lambda\nu}\right) \mathrm{d}\lambda.$$

Representation for the total field can be written in the following form:

$$E_\varphi = \mathrm{e}^{ikp\eta}\frac{\sin\varphi}{\eta} \int\limits_0^{+\infty} \mathrm{e}^{-\lambda/\eta} \Big\{ J_0 \left(\mathrm{e}^{\mathrm{i}\pi/4}\sqrt{2\lambda\nu}\right)$$
$$+ R_0(\lambda) H_0^{(1)} \left(\mathrm{e}^{\mathrm{i}\pi/4}\sqrt{2\lambda\nu}\right) - R_2(\lambda) H_2^{(1)} \left(\mathrm{e}^{\mathrm{i}\pi/4}\sqrt{2\lambda\nu}\right) \Big\} \mathrm{d}\lambda,$$

$$H_\varphi = \mathrm{e}^{ikp\eta}\frac{\cos\varphi}{\eta}\int_0^{+\infty} \mathrm{e}^{-\lambda/\eta}\Big\{J_0\left(\mathrm{e}^{\mathrm{i}\pi/4}\sqrt{2\lambda\nu}\right) + R_0(\lambda)H_0^{(1)}\left(\mathrm{e}^{\mathrm{i}\pi/4}\sqrt{2\lambda\nu}\right) + R_2(\lambda)H_2^{(1)}\left(\mathrm{e}^{\mathrm{i}\pi/4}\sqrt{2\lambda\nu}\right)\Big\}\mathrm{d}\lambda. \tag{5.40}$$

Functions R_0 and R_2 should be chosen such that the boundary conditions on the surface of the cone are satisfied. In the case of the perfect conductor, we have

$$E_\varphi\big|_{\nu=\chi} = 0, \qquad E_\eta\big|_{\nu=\chi} = 0.$$

From the first condition, we get the equation

$$R_0 H_0^{(1)}\left(\mathrm{e}^{\mathrm{i}\pi/4}\sqrt{2\lambda\chi}\right) - R_2 H_2^{(1)}\left(\mathrm{e}^{\mathrm{i}\pi/4}\sqrt{2\lambda\chi}\right) = -J_0\left(\mathrm{e}^{\mathrm{i}\pi/4}\sqrt{2\lambda\chi}\right). \tag{5.41}$$

From the second boundary condition with the use of the formula

$$E_\eta = \left(2\frac{\partial}{\partial\nu} + 1\right) H_\varphi - E_\varphi,$$

we derive the equation

$$R_0\left[\mathrm{e}^{\mathrm{i}\pi/4}\sqrt{2\lambda\chi}\dot{H}_0^{(1)}\left(\mathrm{e}^{\mathrm{i}\pi/4}\sqrt{2\lambda\chi}\right) + H_0^{(1)}\left(\mathrm{e}^{\mathrm{i}\pi/4}\sqrt{2\lambda\chi}\right)\right] + R_2\left[\mathrm{e}^{\mathrm{i}\pi/4}\sqrt{2\lambda\chi}\dot{H}_2^{(1)}\left(\mathrm{e}^{\mathrm{i}\pi/4}\sqrt{2\lambda\chi}\right) + H_2^{(1)}\left(\mathrm{e}^{\mathrm{i}\pi/4}\sqrt{2\lambda\chi}\right)\right] = -\mathrm{e}^{\mathrm{i}\pi/4}\sqrt{2\lambda\chi}\dot{J}_0\left(\mathrm{e}^{\mathrm{i}\pi/4}\sqrt{2\lambda\chi}\right) - J_0\left(E^{\mathrm{i}\pi/4}\sqrt{2\lambda\chi}\right). \tag{5.42}$$

Solving the system of (5.41) and (5.42), we find

$$R_0 = -\frac{\mathrm{i}}{\pi Z(\lambda)}\frac{H_2^{(1)}\left(\mathrm{e}^{\mathrm{i}\pi/4}\sqrt{2\lambda\chi}\right)}{H_0^{(1)}\left(\mathrm{e}^{\mathrm{i}\pi/4}\sqrt{2\lambda\chi}\right)} - \frac{J_0\left(\mathrm{e}^{\mathrm{i}\pi/4}\sqrt{2\lambda\chi}\right)}{H_0^{(1)}\left(\mathrm{e}^{\mathrm{i}\pi/4}\sqrt{2\lambda\chi}\right)}, \qquad R_2 = -\frac{\mathrm{i}}{\pi Z(\lambda)},$$

where

$$Z(\lambda) = \mathrm{e}^{\mathrm{i}\pi/4}\sqrt{\frac{\lambda\chi}{2}}H_1^{(1)}\left(\mathrm{e}^{\mathrm{i}\pi/4}\sqrt{2\lambda\chi}\right) \times \left[H_0^{(1)}\left(\mathrm{e}^{\mathrm{i}\pi/4}\sqrt{2\lambda\chi}\right) - H_2^{(1)}\left(\mathrm{e}^{\mathrm{i}\pi/4}\sqrt{2\lambda\chi}\right)\right].$$

Letting $\nu = \chi$ in formula (5.40), we get the current on the surface of the cone

$$J = \mathrm{e}^{ikp\eta}F_{PEC}\left(2kp\gamma^2\eta\right)\cos\varphi, \tag{5.43}$$

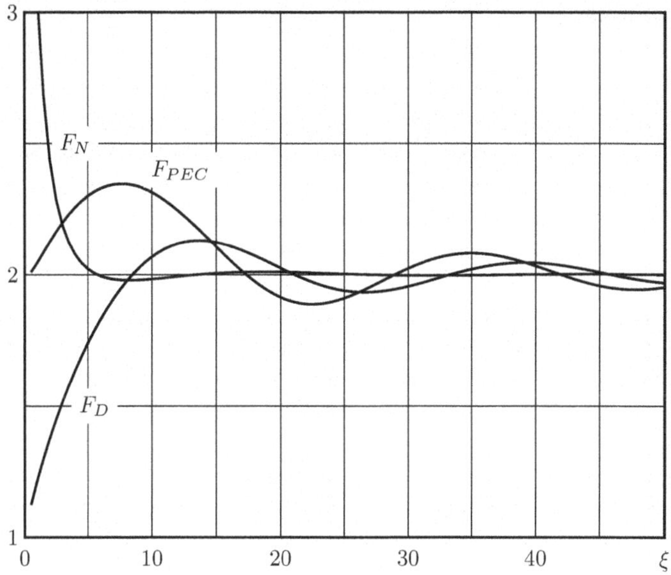

Fig. 5.2 Special functions $F_N(\xi)$, $F_D(\xi)$ and $F_{PEC}(\xi)$

where

$$F_{PEC}(\xi) = -\frac{8}{\pi} e^{i\pi/4} \frac{1}{\xi} \int_0^{+\infty} \exp\left(-\frac{t^2}{\xi}\right) \times$$
$$\times \frac{H_2^{(1)}\left(e^{i\pi/4} t\right)}{H_1^{(1)}\left(e^{i\pi/4} t\right)\left[H_0^{(1)}\left(e^{i\pi/4} t\right) - H_2^{(1)}\left(e^{i\pi/4} t\right)\right]} dt.$$

Absolute values of function $F_{PEC}(\xi)$ are presented in Fig. 5.2.

It is well known that in the general case the currents on the cone are functions of the two parameters, the distance from the tip measured in wavelengths, i.e. kz, and the angle of the cone γ. However, if γ is small, asymptotic approximations (5.37), (5.38) and (5.43) show that in the leading order the field of diffraction by a narrow cone is defined by special functions F_N, F_D or F_{PEC} of only one parameter

$$2kp\gamma^2\eta \approx 2kz\gamma^2.$$

The same remains true in the case of the incidence at a small angle ϑ_0 to the axis of the cone. Namely, the currents are defined not by the usual four parameters z, γ, ϑ_0 and φ, but only by the three parameters $kz\gamma^2$, ϑ_0/γ and φ.

5.8 Conclusion

In this chapter, we considered the problems of diffraction by strongly elongated surfaces of revolution of non-elliptic shape. We mainly restricted ourselves to the analysis in the case of the plane wave incidence along the axis of the body, and to the case of the scalar wave diffraction. However, it can be shown that if the problem of axially incident acoustic wave diffraction allows the leading order term of the asymptotic expansion to be written in explicit form, then in the case of other types of the incident wave (plane wave incident at an angle to the axis, point source fields, etc.) and for the case of diffraction of electromagnetic waves, the leading order terms of the asymptotic expansion can be also written in a similar explicit form. Unfortunately, such explicit representation is possible only in the case of the surfaces of the first and second order, which are the cone or wedge, spheroid or elliptic cylinder, and one- and two-sheeted hyperboloids of revolution or hyperbolic cylinders. For the case of the paraboloid and parabolic cylinder, the asymptotic approach developed in this book does not add anything new, because the boundary-value problems allow variable separationin their initial formulation, that is for the Helmholtz or Maxwell equations in the corresponding paraboloidal or parabolic coordinates.

We considered only the ideal boundary conditions on the surfaces, but problems of diffraction by the surface described with impedance boundary conditions can be solved analogously. Indeed, if the boundary condition is written in the form (see Chap. 1)

$$\left.\frac{\partial u}{\partial n}\right|_{n=0} + ikZu\Big|_{n=0} = 0$$

with $Z = O(1)$, then, since the differentiation with respect to n produces the large parameter kp only in the power one half, the leading order approximation will be the same as in the case of the Dirichlet boundary condition and the impedance will manifest itself only at the next order corrections.

The famous ogival surface cannot be described by the asymptotic approach developed in this book, but one can try to consider a body formed of a truncated spheroid and two cones. Then, for each part of the surface, the asymptotic representations are known (for the case of the spheroid, see Chaps. 3 or 4, and for the cone, see Sect. 5.7), and it remains to do some matching of the fields at the boundaries where the cones are joined to the spheroid. The first step toward this end is done in [11].

It is worth noting also the results of [13] on the numerical evaluation of the parabolic equation solutions in the problems of diffraction by elongated surfaces. However, as compared to the case of explicitly solvable integral equations, the approach of [13] gives a bit worse accuracy of approximation.

We remind also that the solutions constructed in this book can serve as the test examples for the purpose of validation of new numerical solvers, which are being developed presently by many authors.

References

1. M. Abramowitz, I.A. Stegun, Handbook of mathematical functions with formulas, graphs and mathematical tables, National Bureau of Standards. Appl. Math. Ser. **55** (1964)
2. I.V. Andronov, Diffraction by a narrow cone at skew incidence. J. Math. Sci. **226**(6), 695–700 (2017)
3. I.V. Andronov, High-frequency diffraction by a narrow hyperboloid of revolution. Acoust. Phys. **63**(2), 133–140 (2017)
4. I.V. Andronov, D. Bouche, Diffraction by a narrow circular cone as by a strongly elongated body. J. Math. Sci. **185**(4), 517–522 (2012)
5. I.V. Andronov, V.E. Petrov, Diffraction by an impedance strip at almost grazing incidence. IEEE Trans. Antennas Propag. **64**(8), 3565–3572 (2016)
6. V.M. Babich, Diffraction of plane wave by a narrow cone in the case of Dirichlet boundary condition. J. Math. Sci. **86**(3), 2657–2663 (1997)
7. J.J. Bowman, T.B.A. Senior, P.L.E. Uslenghi, *Electromagnetic and Acoustic Scattering by Simple Shapes* (North-Holland Publishing Co., Amsterdam, 1969)
8. I.S. Gradshtein, I.M. Ryzhik, *Table of Integrals, Series and Products* (Elsevier, Academic Press, New York, 2007)
9. E. Kamke, *Differentialgleichungen: L osungsmethoden und L osungen* (Leipzig, 1959)
10. I.V. Komarov, L.I. Ponomarev, *SYu* (Slavyanov, Spheroidal and Coulomb Spheroidal Functions (Science, Moscow, 1976)
11. F. Molinet, I.V. Andronov, Elliptic cylinder with a strongly elongated cross-section: High frequency techniques and function theoretic methods, Chapter 6 in the book, in *Advances in Mathematical Methods for Electromagnetics*, ed. by K. Kobayashi, P.D. Smith (The IET, London, UK, 2020)
12. A.V. Shanin, Diffraction series on a sphere and conical asymptotics, in *Procedings of the International Conference Days on Diffraction* (St.Petersburg, 2011), pp. 88–93
13. A.V. Shanin, A.I. Korolkov, Diffraction by an elongated body of revolution. A boundary integral equation based on the parabolic equation. Wave Motion **85**, 176–190 (2019)
14. V.P. Smyshlyaev, Diffraction by conical surfaces at high frequencies. Wave Motion **12**(4), 329–339 (1990)
15. I.J. Thompson, A.R. Barnett, COULCC: A continued-fraction algorithm for Coulomb functions of complex order with complex arguments. Comput. Phys. Commun. **36**, 363–372 (1985)

Appendix A
Airy and Whittaker Functions

A.1 Airy Functions

Classical asymptotic expansion for the field of diffraction by a smooth convex body is written in terms of the Airy functions. Here, we present some useful formulas concerning these functions.

The standard solutions of the Airy equation

$$Lw(z) \equiv w''(z) - zw(z) = 0$$

are $\mathrm{Ai}(z)$ and $\mathrm{Bi}(z)$ [1]. However, in diffraction theory it is more convenient to use Airy functions in Fock notations

$$v(z) = \sqrt{\pi}\mathrm{Ai}(z),$$

$$w_1(z) = \sqrt{\pi}\Big[\mathrm{Bi}(z) + \mathrm{i}\,\mathrm{Ai}(z)\Big] \tag{A.1}$$

and

$$w_2(z) = \sqrt{\pi}\Big[\mathrm{Bi}(z) - \mathrm{i}\,\mathrm{Ai}(z)\Big].$$

These three solutions are not independent, but satisfy the relation:

$$v(z) = \frac{w_1(z) - w_2(z)}{2\mathrm{i}} = w_3(z).$$

The integral representations for the Airy functions are [3]

$$w_j(z) = \frac{1}{\sqrt{\pi}} \int\limits_{\mathcal{L}_j} \exp\left(tz - \frac{t^3}{3}\right) \mathrm{d}t, \tag{A.2}$$

I. Andronov, *Problems of High Frequency Diffraction by Elongated Bodies*, Springer Series in Optical Sciences 243, https://doi.org/10.1007/978-981-99-1276-6

Table A.1 Zeros of Airy function and its derivative

ℓ	ξ_ℓ	$\dot{\xi}_\ell$
1	2.338107410	1.018792972
2	4.087949444	3.248197582
3	5.520559828	4.820099211
4	6.786708090	6.163307356
5	7.944133587	7.372177255
5(as)	7.942486664	7.374853106

where $\mathcal{L}_j$ are the infinite contours: $\mathcal{L}_1 = (\infty e^{-2\pi i/3}, +\infty)$, $\mathcal{L}_2 = (\infty e^{2\pi i/3}, +\infty)$ and $\mathcal{L}_2 = (\infty e^{-2\pi i/3}, \infty e^{2\pi i/3})$. The integral representation for the Airy function v can be rewritten also in the form

$$v(z) = \frac{1}{\sqrt{\pi}} \int_0^{+\infty} \cos\left(zt + \frac{t^3}{3}\right) dt = \frac{1}{\sqrt{\pi}} \int_{-\infty}^{+\infty} \exp\left(zt + \frac{t^3}{3}\right) dt. \tag{A.3}$$

This formula shows that Airy function $v(z)$ is real for real z. It has an infinite number of zeros $z = -\xi_\ell$ on the negative semiaxis and its derivative $v'(z)$ also has an infinite number of zeros $z = -\dot{\xi}_\ell$. Table A.1 presents the values of ξ_ℓ and $\dot{\xi}_\ell$ for $\ell = 1, 2, \ldots, 5$.

Using the relation [3]

$$w_1(z) = 2e^{i\pi/6} v\left(e^{2\pi i/3} z\right),$$

one can find the solutions of the dispersion equations

$$w_1(\xi) = 0 \quad \text{and} \quad \dot{w}_1(\xi) = 0$$

which appear in the asymptotic representations for the fields in the Fock domain in the case of the ideal boundary conditions. The dispersion equation on the impedance surface

$$\dot{w}_1(\xi) - iq_0 w_1(\xi) = 0$$

is studied in [2].

For the large argument by applying the saddle point method to the integrals (A.2), one gets the asymptotic expansions [3] of the Airy functions

$$v(z) \sim z^{-1/4} \exp\left(-\frac{2}{3} z^{3/2}\right),$$

$$w_{1,2}(z) \sim z^{-1/4} \exp\left(\frac{2}{3} z^{3/2}\right),$$

$$v(-z) \sim z^{-1/4} \sin\left(\frac{2}{3}z^{3/2} + \frac{\pi}{4}\right), \tag{A.4}$$

$$w_1(-z) \sim z^{-1/4} \exp\left(\frac{2i}{3}z^{3/2} + \frac{\pi i}{4}\right), \tag{A.5}$$

$$w_2(-z) \sim z^{-1/4} \exp\left(-\frac{2i}{3}z^{3/2} - \frac{\pi i}{4}\right).$$

Formula (A.4) allows the asymptotic approximations for the zeros ξ_ℓ and $\dot{\xi}_l$ to be found

$$\xi_\ell \sim \left(\frac{3\pi}{8}(4\ell - 1)\right)^{2/3}, \qquad \dot{\xi}_\ell \sim \left(\frac{3\pi}{8}(4\ell - 3)\right)^{2/3}.$$

The last line of Table A.1 presents the asymptotic approximations for the fifth zero of the Airy function $v(z)$ and for the fifth zero of its derivative $v'(z)$.

The Wronskians of the Airy functions do not depend on z and can be found with the use of the asymptotic expansions:

$$\mathcal{W}[w_1, v] = \dot{w}_1(z)v(z) - w_1(z)\dot{v}(z) = 1, \tag{A.6}$$

$$\mathcal{W}[w_1, w_2] = \dot{w}_1(z)w_2(z) - w_1(z)\dot{w}_2(z) = -2i.$$

It is a simple matter to find the solutions to the following inhomogeneous Airy equations with particular right-hand sides

$$\begin{gathered} L\dot{w}_1(z) = w_1(z), \\ Lzw_1(z) = 2\dot{w}_1(z), \\ Lz\dot{w}_1(z) = 3zw_1(z), \\ Lz^2w_1(z) = 4z\dot{w}_1(z) + 2w_1(z). \\ \dots\dots\dots\dots \end{gathered} \tag{A.7}$$

A.2 Whittaker Functions

In this section, we collect some results on the Whittaker functions. Whittaker functions appear in many asymptotic representations in this book, though the required properties of these functions are scattered in the literature. The main results are taken from [1, 4, 9].

A.2.1 Integral Representations and Functional Relations

The Whittaker equation [1] is the special form of the confluent hypergeometric equation (or the Kummer equation)

$$\frac{d^2 w}{dz^2} + \left(-\frac{1}{4} + \frac{\lambda}{z} + \frac{1/4 - \mu^2}{z^2} \right) w = 0. \tag{A.8}$$

This equation has two singular points, one is regular at zero, and the other is irregular at infinity. The singular point at zero is similar to those for cylindrical functions and the singular point at infinity is similar to the one of the Airy equation. Whittaker functions $M_{\lambda,\mu}(z)$ and $W_{\lambda,\mu}(z)$ are the regular and the irregular solutions of equation (A.8). The regular solution $M_{\lambda,\mu}(z)$ is defined by the integral representation [4, formula 9.221]

$$M_{\lambda,\mu}(z) = \frac{z^{\mu+\frac{1}{2}}\Gamma(2\mu+1)}{2^{2\mu}\Gamma\left(\mu+\lambda+\frac{1}{2}\right)\Gamma\left(\mu-\lambda+\frac{1}{2}\right)} \times$$
$$\times \int_{-1}^{1} (1+\sigma)^{\mu-\lambda-\frac{1}{2}}(1-\sigma)^{\mu+\lambda+\frac{1}{2}} e^{\frac{1}{2}z\sigma}\, d\sigma. \tag{A.9}$$

Here, one can take only such values of λ, μ and z for which the integral in the right-hand side converges and no ambiguities are present in the outer multiplier. For the other values of the indices and of the argument z, the solution is obtained by analytic continuation.

The irregular solution is defined by the integral representation [4, formula 9.222]

$$W_{\lambda,\mu}(z) = \frac{z^{\mu+\frac{1}{2}} e^{-z/2}}{\Gamma\left(\mu-\lambda+\frac{1}{2}\right)} \int_{0}^{+\infty} \sigma^{\mu-\lambda-\frac{1}{2}}(1+\sigma)^{\mu+\lambda-\frac{1}{2}} e^{-z\sigma}\, d\sigma, \tag{A.10}$$

which is valid for $\Re(\mu-\lambda) > -\frac{1}{2}$ and the argument of complex z being in the interval $(-\pi, \pi)$. Analytic continuation of function $W_{\lambda,\mu}(z)$ requires the cut on the complex plane of z to be introduced with the branch point at zero.

It is known [9] that function $M_{\lambda,\mu}(z)$ considered as the function of the indices is analytic everywhere except the points $\mu = -\frac{1}{2}, -1, -\frac{3}{2}, -2, \ldots$ where it has simple poles originated by the multiplier $\Gamma(2\mu+1)$. Function $W_{\lambda,\mu}(z)$ is defined for all values of λ and μ.

The Whittaker equation preserves its form if μ is replaced with $-\mu$ and in the case of simultaneous change of z and λ into $-z$ and $-\lambda$. Therefore, functions $M_{\lambda,-\mu}(z)$, $W_{\lambda,-\mu}(z)$, $M_{-\lambda,\mu}(-z)$, $W_{-\lambda,\mu}(-z)$, $M_{-\lambda,-\mu}(-z)$ and $W_{-\lambda,-\mu}(-z)$ are also the solutions of equation (A.8). We present here some relations for these solutions. The following formulas are taken from [4] where they are numbered as 9.232.1, 9.231.2

and 9.233.1 correspondingly

$$W_{\lambda,-\mu}(z) = W_{\lambda,-\mu}(z),$$

$$M_{-\lambda,\mu}(-z) = (-1)^{\mu+1/2} M_{\lambda,\mu}(z), \tag{A.11}$$

$$M_{\mathrm{i}t,\mu}(-\mathrm{i}\nu) = \frac{\Gamma(2\mu+1)}{\Gamma\left(\mu+\frac{1}{2}-\mathrm{i}t\right)} \mathrm{e}^{-\pi t} W_{-\mathrm{i}t,\mu}(\mathrm{i}\nu) + \frac{\Gamma(2\mu+1)}{\Gamma\left(\mu+\frac{1}{2}+\mathrm{i}t\right)} \mathrm{e}^{-\pi t} \mathrm{e}^{-\mathrm{i}\pi(\mu+1/2)} W_{\mathrm{i}t,\mu}(-\mathrm{i}\nu). \tag{A.12}$$

Using (A.12), we get

$$W_{-\mathrm{i}t,\mu}(\mathrm{i}\nu) = \Gamma\left(\mu+\tfrac{1}{2}-\mathrm{i}t\right) \mathrm{e}^{\pi t} \frac{M_{\mathrm{i}t,\mu}(-\mathrm{i}\nu)}{\Gamma(2\mu+1)} - \frac{\Gamma\left(\mu+\frac{1}{2}-\mathrm{i}t\right)}{\Gamma\left(\mu+\frac{1}{2}+\mathrm{i}t\right)} \mathrm{e}^{-\mathrm{i}\pi(\mu+1/2)} W_{\mathrm{i}t,\mu}(-\mathrm{i}\nu). \tag{A.13}$$

We also need the formulas that express the derivatives of the Whittaker functions. These formulas we take from Chap. 8 of [9], which considers Whittaker functions from the point of view of the group theory. In Sect. 3 of chapter 8 of [9] the following expressions for the derivatives of Whittaker functions are given:

$$\dot{W}_{\lambda,\mu}(z) = -\frac{1}{\sqrt{z}} W_{\lambda+\frac{1}{2},\mu-\frac{1}{2}}(z) - \left(\frac{2\mu-1}{2z} - \frac{1}{2}\right) W_{\lambda,\mu}(z), \tag{A.14}$$

$$\dot{W}_{\lambda,\mu}(z) = \frac{\lambda-\mu-\frac{1}{2}}{\sqrt{z}} W_{\lambda-\frac{1}{2},\mu+\frac{1}{2}}(z) + \left(\frac{2\mu+1}{2z} - \frac{1}{2}\right) W_{\lambda,\mu}(z), \tag{A.15}$$

$$\dot{W}_{\lambda,\mu}(z) = \frac{\lambda+\mu-\frac{1}{2}}{\sqrt{z}} W_{\lambda-\frac{1}{2},\mu-\frac{1}{2}}(z) - \left(\frac{2\mu-1}{2z} + \frac{1}{2}\right) W_{\lambda,\mu}(z) \tag{A.16}$$

and

$$\dot{W}_{\lambda,\mu}(z) = -\frac{1}{\sqrt{z}} W_{\lambda+\frac{1}{2},\mu+\frac{1}{2}}(z) + \left(\frac{2\mu+1}{2z} + \frac{1}{2}\right) W_{\lambda,\mu}(z). \tag{A.17}$$

Analogous relations for function M are

$$\dot{M}_{\lambda,\mu}(z) = \frac{2\mu}{\sqrt{z}} M_{\lambda+\frac{1}{2},\mu-\frac{1}{2}}(z) - \left(\frac{2\mu-1}{2z} - \frac{1}{2}\right) M_{\lambda,\mu}(z),$$

$$\dot{M}_{\lambda,\mu}(z) = \frac{\mu - \lambda + \frac{1}{2}}{(2\mu + 1)\sqrt{z}} M_{\lambda - \frac{1}{2}, \mu + \frac{1}{2}}(z) + \left(\frac{2\mu + 1}{2z} - \frac{1}{2}\right) M_{\lambda,\mu}(z),$$

$$\dot{M}_{\lambda,\mu}(z) = \frac{2\mu}{\sqrt{z}} M_{\lambda - \frac{1}{2}, \mu - \frac{1}{2}}(z) - \left(\frac{2\mu - 1}{2z} + \frac{1}{2}\right) M_{\lambda,\mu}(z)$$

and

$$\dot{M}_{\lambda,\mu}(z) = \frac{\mu + \lambda + \frac{1}{2}}{(2\mu + 1)\sqrt{z}} M_{\lambda + \frac{1}{2}, \mu + \frac{1}{2}}(z) + \left(\frac{2\mu + 1}{2z} + \frac{1}{2}\right) M_{\lambda,\mu}(z).$$

Combining (A.14) and (A.16) yields

$$\left[\sqrt{z}\frac{d}{dz} + \frac{\mu - 1}{\sqrt{z}} + \frac{1}{2}\sqrt{z}\right]\left[\sqrt{z}\frac{d}{dz} + \frac{\mu - \frac{1}{2}}{\sqrt{z}} - \frac{1}{2}\sqrt{z}\right] W_{\lambda,\mu}(z) =$$
$$= -\left(\lambda + \mu - \frac{1}{2}\right) W_{\lambda,\mu-1}(z).$$

With the use of Whittaker equation (A.8), this formula reduces to

$$(1 - 2\mu)\dot{W}_{\lambda,\mu}(z) = \left(\frac{(1 - 2\mu)^2}{2z} - \lambda\right) W_{\lambda,\mu}(z)$$
$$+ \left(\lambda + \mu - \frac{1}{2}\right) W_{\lambda,\mu-1}(z). \qquad \text{(A.18)}$$

Combining in a similar manner (A.15) and (A.17) yields

$$(1 + 2\mu)\dot{W}_{\lambda,\mu}(z) = \left(\frac{(1 + 2\mu)^2}{2z} - \lambda\right) W_{\lambda,\mu}(z)$$
$$+ \left(\lambda - \mu - \frac{1}{2}\right) W_{\lambda,\mu+1}(z). \qquad \text{(A.19)}$$

In order to find the Wronskian of Whittaker functions one can use formulas 13.1.32 and 13.1.33 from [1], which express Whittaker functions via Kummer confluent hypergeometric functions M and U as

$$M_{\kappa,\mu}(z) = e^{-z/2} z^{1/2+\mu} M\left(\frac{1}{2} + \mu - \kappa, 1 + 2\mu, z\right), \qquad \text{(A.20)}$$

$$W_{\kappa,\mu}(z) = e^{-z/2} z^{1/2+\mu} U\left(\frac{1}{2} + \mu - \kappa, 1 + 2\mu, z\right),$$

and use formula 13.1.22 from [1] for the Wronskian of Kummer functions

$$M(a,b,z)\frac{\mathrm{d}U(a,b,z)}{\mathrm{d}z} - U(a,b,z)\frac{\mathrm{d}M(a,b,z)}{\mathrm{d}z} = -\frac{\Gamma(b)}{\Gamma(a)}z^{-b}\mathrm{e}^{z}.$$

That yields

$$\mathcal{W}[M,W] = \dot{M}_{\kappa,\mu}(z)W_{\kappa,\mu}(z) - M_{\kappa,\mu}(z)\dot{W}_{\kappa,\mu}(z) = \frac{\Gamma(2\mu+1)}{\Gamma\left(\frac{1}{2}+\mu-\kappa\right)}. \tag{A.21}$$

For the regular Whittaker function, we also need the representation for the small values of its argument

$$M_{it,\mu}(z) \sim z^{1/2+\mu},\ |z| \to 0. \tag{A.22}$$

This formula follows from (A.20) and the series for Kummer function M given by formula 13.1.2 from [1].

Some results concerning Whittaker functions are formulated in terms of Coulomb wave functions F and H^{+}.

A.2.2 The Coulomb Wave Functions

Rewriting the formulas in terms of the Coulomb wave function can be also useful for performing computations. To compute the values of functions F and H^{+}, one can use the code coulcc explained in [8] and available in FORTRAN at https://github.com/fnevgeny/cfac/blob/master/coul/coulcc.f. This code accepts complex values of the argument and the indices of Coulomb wave functions.

As is known (see, for example, [1]), Coulomb wave functions $F_L(t,z)$, $G_L(t,z)$ and $H_L^{\pm}(t,z) = G_L(t,z) \pm \mathrm{i}F_L(t,z)$ are the solutions of the equation

$$w''(z) + \left[1 - \frac{2t}{z} - \frac{L(L+1)}{z^2}\right]w(z) = 0$$

and therefore differ from the Whittaker functions only by the shift in the index and by normalization. Though Chap. 14 of [1] considers L to be a nonnegative integer, most formulas presented there remain valid for negative and fractional values of L as well, since the Coulomb wave functions are related to the Whittaker functions which are analytic with respect to both indices. The relations between the Coulomb wave functions and the Kummer functions are given by formulas 13.6.8 or 14.1.3 and formula 14.1.7 from [1]. However, the absolute value of gamma-function presented in the last formula should be interpreted as

$$|\Gamma(L+1+\mathrm{i}\eta)| = \sqrt{\Gamma(L+1+\mathrm{i}\eta)\Gamma(L+1-\mathrm{i}\eta)}.$$

From the above-mentioned formulas from [1], one can derive the following relations:

$$M_{\mathrm{i}t,\mu}(\mathrm{i}z) = \frac{2\Gamma(2\mu+1)\exp\left(\mathrm{i}\frac{\pi}{4}(2\mu+1)+\frac{\pi}{2}t\right)}{\sqrt{\Gamma\left(\mu+\frac{1}{2}+\mathrm{i}t\right)\Gamma\left(\mu+\frac{1}{2}-\mathrm{i}t\right)}} F_{\mu-\frac{1}{2}}\left(t,\frac{z}{2}\right)$$

$$M_{\mathrm{i}t,\mu}(-\mathrm{i}z) = \frac{2\Gamma(2\mu+1)\exp\left(-\mathrm{i}\frac{\pi}{4}(2\mu+1)-\frac{\pi}{2}t\right)}{\sqrt{\Gamma\left(\mu+\frac{1}{2}+\mathrm{i}t\right)\Gamma\left(\mu+\frac{1}{2}-\mathrm{i}t\right)}} F_{\mu-\frac{1}{2}}\left(-t,\frac{z}{2}\right)$$

$$W_{\mathrm{i}t,\mu}(-\mathrm{i}z) = \exp\left(\mathrm{i}\frac{\pi}{4}(2\mu-1)+\frac{\pi}{2}t\right)\frac{\sqrt{\Gamma\left(\mu+\frac{1}{2}+\mathrm{i}t\right)}}{\sqrt{\Gamma\left(\mu+\frac{1}{2}-\mathrm{i}t\right)}}\times$$

$$\times H^{+}_{\mu-\frac{1}{2}}\left(-t,\frac{z}{2}\right).$$

A.2.3 Asymptotic Expansions

For the irregular Whittaker functions, the asymptotics at $|z| \to \infty$ is given by formula 9.227 from [4]. Its leading order term is

$$W_{\lambda,\mu}(z) \sim \mathrm{e}^{-z/2} z^{\lambda}. \tag{A.23}$$

The asymptotics of the regular Whittaker function of large argument

$$M_{\mathrm{i}t,\ell}(z) \sim \exp\left(-\frac{z}{2}+\mathrm{i}\pi\left(\ell+\frac{1}{2}-\mathrm{i}t\right)\right)\frac{\Gamma(2\ell+1)}{\Gamma\left(\ell+\frac{1}{2}+\mathrm{i}t\right)}z^{\mathrm{i}t} + \exp\left(\frac{z}{2}\right)\frac{\Gamma(2\ell+1)}{\Gamma\left(\ell+\frac{1}{2}-\mathrm{i}t\right)}z^{-\mathrm{i}t} \tag{A.24}$$

can be found by substituting from (A.23) into functional relation (A.12). Asymptotic formula (A.24) is valid for $\arg(z) > -\frac{\pi}{2}$.

The asymptotic formulas for the Whittaker functions with both the first index and the argument being large can be derived by using the relations between the Whittaker functions and the Coulomb wave functions and the asymptotic formula 14.6.12 from [1]. It reads

$$H^{+}_{L}(t,z) = F_L(t,r) + \mathrm{i}\,G_L(t,r) \sim \left(\frac{z_L}{1+L(L+1)/z_L^2}\right)^{1/6} w_1(\zeta) \tag{A.25}$$

and is valid for $t \gg 1$, $z \sim 2t$. Here, Ai and Bi are the Airy functions [1] and

$$z_L = t + \sqrt{t^2 + L(L+1)}, \qquad \zeta = (z_L - r)\left(\frac{1}{r_L} + \frac{L(L+1)}{r_L^3}\right)^{1/3}.$$

By substituting $z = \chi/2$ in (A.25) and neglecting smaller order corrections for the case of finite values of L, we get the approximations

$$z_L \approx 2t, \qquad \zeta \approx \xi = \left(2t - \frac{\chi}{2}\right)(2t)^{-1/3}.$$

Finally using Airy function w_1 in Fock notations (A.1), we get

$$H_L^+(t, \chi/2) \sim (2t)^{1/6} w_1(\xi), \qquad \dot{H}_L^+(t, \chi/2) \sim -(2t)^{-1/6} \dot{w}_1(\xi). \tag{A.26}$$

A.3 Solutions of Dispersion Equations

The integrals representing asymptotic approximations for the solutions in the problems of diffraction by an elliptic cylinder with strongly elongated cross-section and by strongly elongated spheroids can be computed by using the residue theorem. This requires the zeros of the dispersion equations to be studied, which we do in this section.

In the case of the acoustic wave diffraction by the elliptic cylinder, the dispersion equations are

$$W_{\mathrm{i}t,1/4}(-\mathrm{i}\chi) = 0 \quad \text{and} \quad 4\mathrm{i}\chi \dot{W}_{\mathrm{i}t,1/4}(-\mathrm{i}\chi) + W_{\mathrm{i}t,1/4}(-\mathrm{i}\chi) = 0 \tag{A.27}$$

correspondingly in the case of the absolutely soft and the absolutely hard surfaces. In the problem of diffraction by the spheroid, the corresponding dispersion equations are

$$W_{\mathrm{i}t,n/2}(-\mathrm{i}\chi) = 0 \quad \text{and} \quad 2\mathrm{i}\chi \dot{W}_{\mathrm{i}t,n/2}(-\mathrm{i}\chi) + W_{\mathrm{i}t,n/2}(-\mathrm{i}\chi) = 0 \tag{A.28}$$

with $n = 0, 1, 2, \ldots$ being the index of Fourier harmonics. And in the case of the electromagnetic wave diffraction by strongly elongated spheroids, we deal with the dispersion equation (4.24)

$$\begin{aligned} Z_n(t) = {} & W_{\mathrm{i}t,(n-1)/2}(-\mathrm{i}\chi)\dot{W}_{\mathrm{i}t,(n+1)/2}(-\mathrm{i}\chi) \\ & + \dot{W}_{\mathrm{i}t,(n-1)/2}(-\mathrm{i}\chi)W_{\mathrm{i}t,(n+1)/2}(-\mathrm{i}\chi) = 0, \end{aligned} \tag{A.29}$$

which can be also rewritten as (4.25)

$$\left(n - 2\mathrm{i}t\right)W^2_{\mathrm{i}t,(n-1)/2}(-\mathrm{i}\chi) + \left(n + 2\mathrm{i}t\right)W^2_{\mathrm{i}t,(n+1)/2}(-\mathrm{i}\chi) = 0.$$

All these equations can be solved only numerically. We base our computations on the Newton iterations method, but it needs the initial approximation. This approximation can be obtained by considering parameter χ to be asymptotically large. From asymptotic representation (A.26), one can conclude that the solutions of the above dispersion equations in the case of large χ can be approximated by the following expressions:

$$t_{n\ell} = -\frac{\chi}{4} + \frac{\xi_\ell}{2}\left(\frac{\chi}{2}\right)^{1/3} \mathrm{e}^{-2\pi\mathrm{i}/3}, \tag{A.30}$$

$$t_{n\ell} = -\frac{\chi}{4} + \frac{\dot{\xi}_\ell}{2}\left(\frac{\chi}{2}\right)^{1/3} \mathrm{e}^{-2\pi\mathrm{i}/3}, \tag{A.31}$$

where ξ_ℓ and $\dot{\xi}_\ell$ are the zeros of the Airy function $v(-\xi)$ and its derivative $\dot{v}(-\xi)$ (see Sect. A.1). The approximation (A.30) is valid for the zeros of the dispersion equations corresponding to the Dirichlet boundary condition satisfied by the field on the surface, while (A.31) approximates the solutions of the dispersion equations corresponding to the problems with the Neumann boundary condition. In the case of the dispersion equation for the electromagnetic waves, both approximations are valid. That is, in this case there are zeros approximated by (A.30) and there are zeros approximated by (A.31).

It is worth noting that the index n is not presented on the right-hand sides of (A.30) and (A.31), but when n increases the approximations become less accurate, and one needs to increase the elongation parameter χ in order to maintain accuracy.

We define functions $t_{n\ell}(\chi)$ as the solutions of a particular dispersion equation for given n and χ. First, by setting χ sufficiently large and using the approximations (A.30) or/and (A.31), one can start the Newton iteration procedure and find "exact" values of $t_{n\ell}(\chi)$ for a given large value of χ. Then decreasing χ, one uses the already found values of the zeros as the initial approximations and again starts the Newtonian iterations to find the "exact" values. By this procedure, decreasing step by step the values of χ, one can trace the solutions up to any desired value of the elongation parameter. Results of these computations are presented in Figs. A.1, A.2 and A.3.

Blue curves are for the case of the soft surface and red curves are for the case of the hard surface. The values of χ are marked by bullets on the curves. The leftmost bullet on each curve corresponds to $\chi = 10$, then next to $\chi = 5$, then to $\chi = 2, 1, \frac{1}{2}, \frac{1}{4}, \frac{1}{10}, \frac{1}{20}$ and $\frac{1}{40}$.

Computations show that when χ decreases, the solutions $t_{n\ell}$ trace the curves moving to the right on the complex plane of t. Most of the solutions approach the points $-\mathrm{i}\ell/2$, when $\chi \to 0$, but some tend to infinity. One can also note that in the case of electromagnetic waves, one of the solutions of the dispersion equation for the 0-th order harmonics approaches the real axis of t when χ decreases. In the case of $n = 0$, the dispersion equation reduces to two equations, either

$$W_{\mathrm{i}t,1/2}(-\mathrm{i}\chi) = 0 \qquad \text{or} \qquad \dot{W}_{\mathrm{i}t,1/2}(-\mathrm{i}\chi) = 0.$$

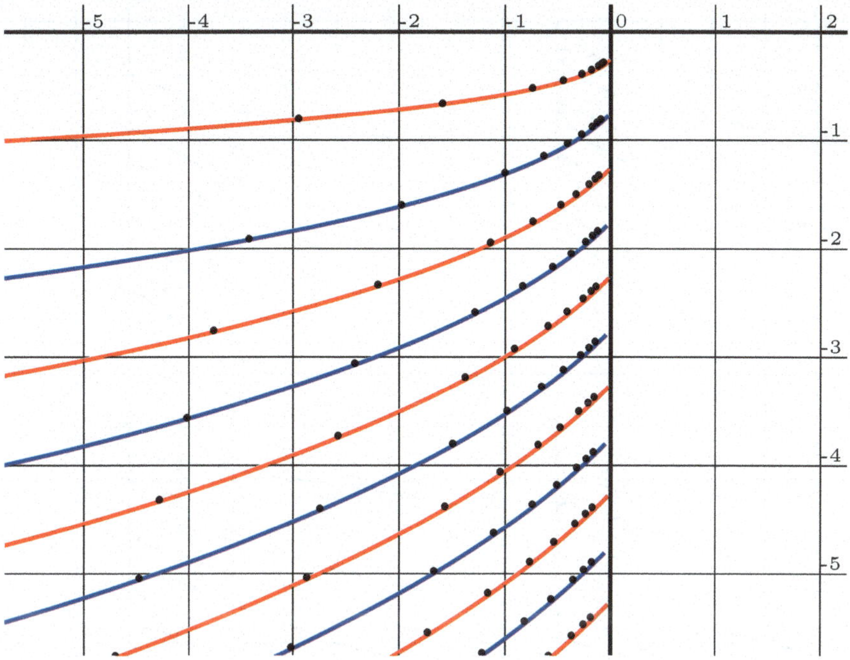

Fig. A.1 The curves on the complex plane of t representing the solutions $t_{1/4,\ell}(\chi)$ of the dispersion equations (A.27)

The solution t_{0*} which approaches the real axis corresponds to the second of these equations, and the corresponding term is presented only in the case of TM polarizations as can be seen from (4.29).

A.4 The Integral Transform

In Chaps. 2, 3 and 4 of this book, we used the following integral transform:

$$\hat{f}(\eta) = \int_{-\infty}^{+\infty} \left(\frac{1-\eta}{1+\eta}\right)^{it} f(t)\mathrm{d}t, \qquad -1 < \eta < 1. \tag{A.32}$$

In a bit different form, this transform appears in [5, 6]. The transform (A.32) reduces to inverse Mellin transform if we express variable η via new variable s with the formula

$$\eta = \frac{s-1}{s+1}.$$

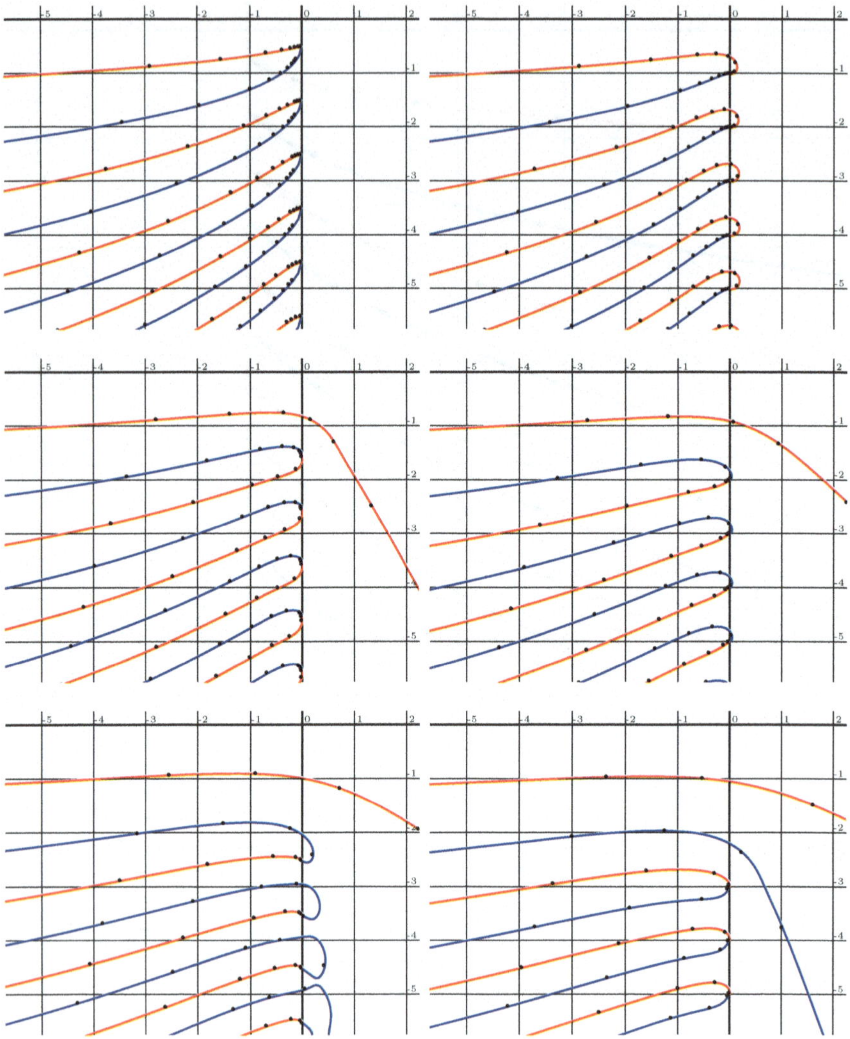

Fig. A.2 The curves on the complex plane of t representing the solutions $t_{n/2,\ell}(\chi)$ of the dispersion equations (A.28) for from $n = 0$ (top-left) to $n = 5$ (bottom-right)

Then

$$\frac{1-\eta}{1+\eta} = \frac{1}{s}$$

and the transform (A.32) reduces to

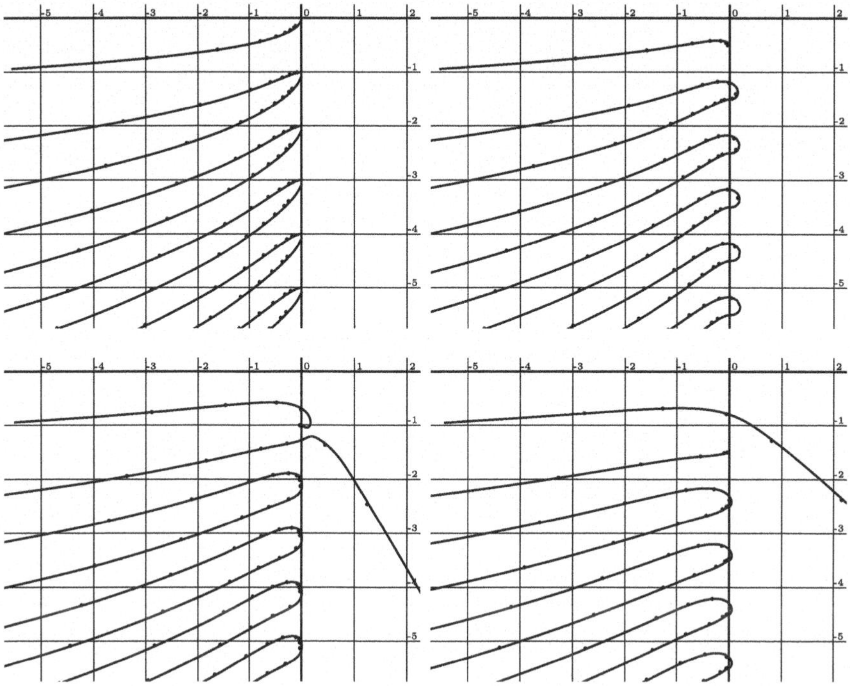

Fig. A.3 The curves on the complex plane of t representing the solutions $t_{n/2,\ell}(\chi)$ of the dispersion equations (A.29) for from $n = 0$ (top-left) to $n = 3$ (bottom-right)

$$\hat{f}(\eta) = \int_{-\mathrm{i}\infty}^{+\mathrm{i}\infty} s^{-\mathrm{i}t} f(t)\mathrm{d}t.$$

Its inversion is evidently given by the direct Mellin transform

$$f(t) = \frac{1}{2\pi} \int_{0}^{+\infty} \hat{f}\left(\frac{s-1}{s+1}\right) s^{\mathrm{i}t-1}\mathrm{d}s.$$

Returning to variable η yields

$$f(t) = \frac{1}{\pi} \int_{-1}^{1} \left(\frac{1+\eta}{1-\eta}\right)^{\mathrm{i}t} \frac{\hat{f}(\eta)}{1-\eta^2}\mathrm{d}\eta. \tag{A.33}$$

When dealing with diffraction by a narrow hyperboloid in Chap. 5, we need to consider the similar integral transform

$$\hat{f}(\eta) = \int_{-\infty}^{+\infty} \left(\frac{\eta+1}{\eta-1}\right)^{it} f(t)\mathrm{d}t, \qquad |\eta| > 1. \tag{A.34}$$

In that case, we set

$$s = \frac{\eta+1}{\eta-1} = 1 + \frac{2}{\eta-1}$$

and get the inversion

$$\hat{f}(t) = \frac{1}{2\pi} \int_{0}^{+\infty} f\left(\frac{s+1}{s-1}\right) s^{-it-1}\mathrm{d}s.$$

When rewriting this integral with respect to variable η, we split it into two parts. The part $s \in [0, 1]$ transforms into $\eta \in (-\infty, -1]$ and the part $s \in [1, +\infty)$ transforms into $\eta \in [1, +\infty)$, which yields

$$\hat{f}(t) = \frac{1}{\pi} \int_{(-\infty,-1]\cup]1,+\infty)} F(\eta) \left(\frac{\eta+1}{\eta-1}\right)^{-it-1} \frac{\mathrm{d}\eta}{(\eta-1)^2}. \tag{A.35}$$

A.5 Integrals Containing Whittaker Functions

In this section, we collect the integral identities involving Whittaker functions. These identities appear in the book when deriving integral representations for the incident fields.

The first identity is taken from Chap. 2 and expresses the solution of integral equation (2.18) in explicit form. It can be written as follows:

$$\int_{-\infty}^{+\infty} \left(\frac{1-\eta}{1+\eta}\right)^{it} \Gamma\left(\frac{1}{4}+\mathrm{i}t\right) \Gamma\left(\frac{1}{4}-\mathrm{i}t\right) M_{\mathrm{i}t,-1/4}(\mathrm{i}a) M_{\mathrm{i}t,-1/4}(-\mathrm{i}b)\mathrm{d}t =$$
$$= \sqrt{2}\pi^{3/2}(1-\eta^2)^{1/4}(ab)^{1/4} \exp\left(\frac{\mathrm{i}}{2}(b-a)\eta\right) \cos\left(\sqrt{1-\eta^2}\sqrt{ab}\right). \tag{A.36}$$

Solution (2.22) gives another identity

$$\int_{-\infty}^{+\infty} \left(\frac{1-\eta}{1+\eta}\right)^{it} \Gamma\left(\frac{3}{4}+it\right)\Gamma\left(\frac{3}{4}-it\right) M_{it,1/4}(ia)M_{it,1/4}(-ib)dt =$$

$$= \left(\frac{\pi}{2}\right)^{3/2} (1-\eta^2)^{1/4}(ab)^{1/4} \exp\left(\frac{i}{2}(b-a)\eta\right) \sin\left(\sqrt{1-\eta^2}\sqrt{ab}\right). \quad (A.37)$$

For the case of the line source, somewhat more complicated identities appear, namely

$$\int_{-\infty}^{+\infty} \left(\frac{\zeta-1}{\zeta+1}\frac{1-\eta}{1+\eta}\right)^{it} \Gamma\left(\frac{1}{4}+it\right)\Gamma\left(\frac{1}{4}-it\right) M_{it,-1/4}(ia)M_{it,-1/4}(-ib)dt =$$

$$= \sqrt{2}\pi^{3/2}(ab)^{1/4}\frac{(1-\eta^2)^{1/4}(\zeta^2-1)^{1/4}}{\sqrt{\eta+\zeta}} \exp\left(\frac{i}{2}(b-a)\frac{1+\eta\zeta}{\eta+\zeta}\right)\times$$

$$\times \cos\left(\frac{\sqrt{ab}\sqrt{\zeta^2-1}\sqrt{1-\eta^2}}{\eta+\zeta}\right), \quad (A.38)$$

$$\int_{-\infty}^{+\infty} \left(\frac{\zeta-1}{\zeta+1}\frac{1-\eta}{1+\eta}\right)^{it} \Gamma\left(\frac{3}{4}+it\right)\Gamma\left(\frac{3}{4}-it\right) M_{it,1/4}(ia)M_{it,1/4}(-ib)dt =$$

$$= \left(\frac{\pi}{2}\right)^{3/2}(ab)^{1/4}\frac{(1-\eta^2)^{1/4}(\zeta^2-1)^{1/4}}{\sqrt{\eta+\zeta}} \exp\left(\frac{i}{2}(b-a)\frac{1+\eta\zeta}{\eta+\zeta}\right)\times$$

$$\times \sin\left(\frac{\sqrt{ab}\sqrt{\zeta^2-1}\sqrt{1-\eta^2}}{\eta+\zeta}\right). \quad (A.39)$$

Letting $\zeta \to +\infty$ reduces (A.38) and (A.39) to (A.36) and (A.37).

In Chap. 3, the solution of equation (3.23) is constructed. It is given by (3.28). Combining these formulas, one can write the following identities:

$$\int_{-\infty}^{+\infty} \left(\frac{1-\eta}{1+\eta}\right)^{it} \Gamma\left(\frac{n+1}{2}+it\right)\Gamma\left(\frac{n+1}{2}-it\right) M_{it,n/2}(ia)M_{it,n/2}(-ib)dt =$$

$$= \pi(n!)^2\sqrt{ab}\sqrt{1-\eta^2}\exp\left(\frac{i}{2}(b-a)\eta\right) J_n\left(\sqrt{ab}\sqrt{1-\eta^2}\right). \quad (A.40)$$

Here, we replaced α^2 with a and $\chi\tau$ with b and expressed the gamma-function of the integer argument via the factorial.

In view of the expressions for the Bessel functions of semi-integer index

$$J_{-1/2}(z) = \sqrt{\frac{2}{\pi z}} \cos(z), \quad J_{1/2}(z) = \sqrt{\frac{2}{\pi z}} \sin(z),$$

formulas (A.36) and (A.37) follow from (A.40) by setting $n = -1/2$ and $n = 1/2$ correspondingly. Probably, formulas (A.40) are valid not only for integer values of index n but also for any such n for which the integral converges.

In the case of the point source field diffraction by the strongly elongated spheroid, the following identity is derived:

$$\int\limits_{-\infty}^{+\infty} \left(\frac{\zeta - 1}{\zeta + 1} \frac{1 - \eta}{1 + \eta} \right)^{it} \Gamma\left(\frac{n+1}{2} + \mathrm{i}t \right) \Gamma\left(\frac{n+1}{2} - \mathrm{i}t \right) M_{\mathrm{i}t,n/2}(\mathrm{i}a) \times$$
$$\times M_{\mathrm{i}t,n/2}(-\mathrm{i}b)\mathrm{d}t = \pi (n!)^2 \sqrt{ab} \frac{\sqrt{1-\eta^2}\sqrt{\zeta^2 - 1}}{\eta + \zeta} \times$$
$$\times \exp\left(\frac{\mathrm{i}}{2}(b - a) \frac{1 + \eta\zeta}{\eta + \zeta} \right) J_n \left(\sqrt{ab} \frac{\sqrt{1-\eta^2}\sqrt{\zeta^2-1}}{\eta + \zeta} \right). \quad \text{(A.41)}$$

Again, by setting $n = -1/2$ and $n = 1/2$ formula (A.41) reduces to (A.38) and (A.39).

One more identity is derived in Chap. 4. The right-hand side (4.36) of the integral equation corresponds to the solution given by (4.44). This results in the identity

$$\int\limits_{-\infty}^{+\infty} \left(\frac{\zeta - 1}{\zeta + 1} \frac{1 - \eta}{1 + \eta} \right)^{it} \Gamma\left(\tfrac{n}{2} + \mathrm{i}t \right) \Gamma\left(\tfrac{n}{2} - \mathrm{i}t \right) \times$$
$$\times \left[\frac{n^2 + 4t^2}{2n(n+1)} M_{\mathrm{i}t, \frac{n+1}{2}}(\mathrm{i}a) + (n\zeta - 2\mathrm{i}t) M_{\mathrm{i}t, \frac{n-1}{2}}(\mathrm{i}a) \right] M_{\mathrm{i}t, \frac{n}{2}}(-\mathrm{i}b)\mathrm{d}t =$$
$$= \pi n!(n-1)! \frac{\zeta^2 - 1}{(\eta + \zeta)^2} \exp\left(\frac{\mathrm{i}}{2}(b - a) \frac{1 + \zeta\eta}{\zeta + \eta} \right) \times$$
$$\times \left[\sqrt{a}\sqrt{\zeta^2 - 1} J_{n-1}(\psi) + \mathrm{i}\sqrt{b}\sqrt{1 - \eta^2} J_n(\psi) \right],$$

where

$$\psi = \frac{\sqrt{ab}\sqrt{\eta^2 - 1}\sqrt{1 - \eta^2}}{\zeta + \eta}.$$

Some of the identities presented in this section can be found in [7], but the others seem to be new.

References

1. M. Abramowitz, I.A. Stegun, *Handbook of Mathematical Functions with Formulas, Graphs and Mathematical Tables*, National Bureau of Standards, Applied Mathematics Series 55 (1964)
2. I.V. Andronov, D. Bouche, On the degeneration of creeping waves in a vicinity of critical values of the impedance. Wave Motion **45**, 400–411 (2008)
3. V.M. Babich, B.S. Buldyrev, *Short-Wavelength Diffraction Theory: Asymptotic Methods*, Springer Series on Wave Phenomena, vol. 4 (Springer, 1991), 445 p.
4. I.S. Gradshtein, I.M. Ryzhik, *Table of Integrals, Series and Products* (Elsevier, Academic Press, New York, 2007)
5. V.E. Petrov, Generalized singular Tricomi equation as an equation of convolution. Doklady Math. **74**, 901–905 (2006)
6. V.E. Petrov, Problems of Mathematical Analysis, Collection of works, No 31 (2005)
7. A.P. Prudnikov, Yu.A. Brychkov, O.I. Marichev, Integrals and Series, v. 2 (New York Gordon & Breach, 1986)
8. I.J. Thompson, A.R. Barnett, COULCC: a continued-fraction algorithm for Coulomb functions of complex order with complex arguments. Comput. Phys. Commun. **36**, 363–372 (1985)
9. N.Ya. Vilenkin, Special functions and group representation theory, Moscow, Science (1965)

Index

I. Andronov, *Problems of High Frequency Diffraction by Elongated Bodies*, Springer Series in Optical Sciences 243, https://doi.org/10.1007/978-981-99-1276-6

The manufacturer's authorised representative in the EU is Springer Nature Customer Service Centre GmbH, Europaplatz 3, 69115 Heidelberg, Germany. If you have any concerns regarding our products, please contact ProductSafety@springernature.com

Printed and bound by CPI Group (UK) Ltd, Croydon, CR0 4YY
01/10/2026
02226183-0002